AMC 10 PREPARATION

AMC 10 PREPARATION

Roman Kvasov

CONTENTS

Dedicated to my sister Anna

INTRODUCTION

Every mathematics competition has its unique format, rules, and most importantly - the list of topics it covers. Mastering these topics and techniques is crucial for outstanding performance. This book presents the most popular ideas used to solve problems from the **AMC 10 (American Mathematics Contest)**. It also includes 180 practice problems in AMC 10 format, with full solutions.

The book is structured into 60 chapters, each covering a single topic or method. Each chapter begins with a brief overview of the main idea, followed by practice problems. Some of the problems in the book are well-known, while others are modified versions of existing problems. The majority of the problems, however, are original and come from the authors' personal archive, accumulated over 15 years of math olympiad coaching.

Even though the book covers basic topics in school algebra, geometry, and number theory, it is not intended to replace a textbook. Instead, it is a well-organized, extensive review of the types of problems that appear on AMC 10 competitions year after year. Therefore, for many, it can serve as an excellent introduction to the world of AMC 10. Others may find this book useful in the weeks leading up to the competition, when they want to deepen their knowledge and improve their skills. The book can also serve as a good indicator of which topics should be practiced more extensively.

It is important to note that preparing for any mathematics competition is a long journey that requires a significant amount of time and effort. In addition to understanding
the theory, students should also include a substantial amount of practice in their daily
preparation. Upon completing this book, it is highly recommended to work through
all the past AMC 10 exams, which are widely available online. This will complete
your preparation and boost your confidence.

The author wishes you the best of luck on your math olympiad journey and hopes
you will enjoy the book.

Roman Kvasov, Ph.D.

CHAPTER 1

PERCENTS

A **percent** (usually denoted using the symbol %) is a number written as a fraction with the denominator 100. For example, 27% represents a fraction

$$\frac{27}{100}$$

In order to find a percent of a number, we multiply this number by the fraction representing the percent. For example, 25% of 40 can be found as

$$\frac{25}{100} \cdot 40 = 10$$

In order to find which percent of the total represents a given number, we divide the number by the total and multiply it by 100%. For example, 25 represents the following percent of 40

$$\frac{25}{40} \cdot 100\% = 62.5\%$$

Let us now consider several examples where the percents are used.

Problem 1

Nicolas took two exams. He got 20% of the problems wrong on the first exam and 60% of the problems right on the second exam. Find the fraction of the problems that Nicolas did right if the first exam had 40 problems and the second exam had 30 problems.

(A) $\frac{9}{70}$ **(B)** $\frac{9}{35}$ **(C)** $\frac{4}{5}$ **(D)** $\frac{5}{7}$ **(E)** $\frac{7}{10}$

Solution

Notice that Nicolas got 80% of the problems right on the first exam. Therefore, the number of correct answers on the first exam is

$$\frac{80}{100} \cdot 40 = 32$$

The number of correct answers on the second exam is

$$\frac{60}{100} \cdot 30 = 18$$

Therefore, the total number of correct answers is $32 + 18 = 50$, while the total number of problems is $30 + 40 = 70$. The fraction of the correct answers is

$$\frac{50}{70} = \frac{5}{7}$$

and the right answer is $\boxed{\textbf{(D)} \ \frac{5}{7}}$

Problem 2

Adhish went to a store that offers a 10% discount on its shirts priced at $\$20$ each. He is told that if he buys two shirts, then he will be given another 20% discount on the reduced price of the second shirt. Find the price of the two shirts.

(A) 24.0 **(B)** 28.6 **(C)** 30.2 **(D)** 32.4 **(E)** 34.0

Solution

The reduced price of a shirt is

$$20 - \frac{10}{100} \cdot 20 = 18$$

After the additional 20% discount the price of the second shirt is

$$18 - \frac{20}{100} \cdot 18 = 14.40$$

Therefore, the total price of the two shirts is

$$18 + 14.4 = 32.4$$

and the right answer is $\boxed{\textbf{(D)}\ 32.4}$

Problem 3

The price of a set of plates was decreased by 10% and then increased by 10%. The difference between the original and the final prices is equal to $\$3.50$. What was the original price?

(A) 200   **(B)** 250   **(C)** 300   **(D)** 350   **(E)** 400

Solution

Let the original price be x. The decrease by 10% results in a price

$$x - \frac{10}{100} \cdot x = \frac{90x}{100}$$

The increase by 10% results in a new price

$$\frac{90x}{100} + \frac{10}{100} \cdot \frac{90x}{100} = \frac{90x}{100} + \frac{9x}{100} = \frac{99x}{100}$$

The difference between the original and final prices, therefore, is

$$x - \frac{99x}{100} = \frac{x}{100}$$

From here we have an equation

$$\frac{x}{100} = 3.50$$

which implies that $x = 350$ and the right answer is $\boxed{\textbf{(D)}\ 350}$

CHAPTER 2

RATIOS

A **ratio** describes the relation between the quantities and shows how many times one number is contained in another number. For example, if the ratio of apples to oranges in a box is given as 2 : 3, this means that for every two apples in the box there are three oranges.

Let us now consider several examples where the ratios are used.

Problem 1

The three angles of a triangle satisfy the ratio $5 : 6 : 7$. Which of the following represents the smallest angle?

(A) $45°$ **(B)** $50°$ **(C)** $75°$ **(D)** $80°$ **(E)** $90°$

Solution

Let the angles of the triangle be $5x$, $6x$ and $7x$. Since the sum of all angles in any triangle is $180°$, then we have the following equation

$$5x + 6x + 7x = 180°$$

This equation can be solved by isolating x

$$5x + 6x + 7x = 180°$$
$$18x = 180°$$
$$x = 10°$$

The smallest angle of the triangle can be found as $5x$

$$5x = 5 \cdot 10° = 50°$$

and the right answer is $\boxed{\textbf{(B) } 50°}$

Problem 2

The ratio of gold to silver medals is $1 : 2$ and the ratio of gold to bronze medals is $1 : 3$. It is known that there are 180 more students who obtained a bronze medal than those who obtained a gold medal. Find the total number of students who obtained a medal.

(A) 240   **(B)** 320   **(C)** 360   **(D)** 480   **(E)** 540

Solution

Let x be the number of students who obtained a gold medal. Then $2x$ represents the number of students who obtained a silver medal and $3x$ represents the number of students who obtained a bronze medal. However, we know that there are 180 more students who obtained a bronze medal than those who obtained a gold medal. This implies the following equation

$$x + 180 = 3x$$

This equation can be solved by isolating x

$$x + 180 = 3x$$
$$180 = 2x$$
$$90 = x$$

The total number of students who obtained a medal can be found as

$$x + 2x + 3x = 6x$$

which equals $6 \cdot 90 = 540$ and the right answer is $\boxed{\textbf{(E)} \ 540}$

Problem 3

The ratio $k : k^2 : (2k - 1)$ represents the number of juice, water and soda bottles sold at a store. Which is closes to the percent of the items that represent soda bottles if there are 56 more water bottles than the juice bottles?

(A) 15% **(B)** 17% **(C)** 19% **(D)** 21% **(E)** 23%

Solution

Since there are 56 more water bottles than the juice bottles, then we have

$$k^2 = k + 56$$
$$k^2 - k - 56 = 0$$
$$(k - 8)(k + 7) = 0$$

and, therefore, $k = 8$.

From here we have $k^2 = (8)^2 = 64$ water bottles and $2k - 1 = 2(8) - 1 = 15$ soda bottles. The total number of bottles is

$$8 + 64 + 15 = 87$$

The percent of the soda bottles is

$$\frac{15}{87} \cdot 100\% \approx 17.24\%$$

and the right answer is $\boxed{\textbf{(B)} \ 17\%}$

CHAPTER 3

PROPORTIONS

A **proportion** is an equality of two fractions (or two ratios). For example

$$\frac{12}{15} = \frac{8}{10}$$

A proportion

$$\frac{a}{b} = \frac{c}{d}$$

can be cross-multiplied to obtain an equivalent equality

$$a \cdot d = b \cdot c$$

Many quantities are known to be proportional, which makes the proportion a common topic on AMC 10 competitions.

Let us consider several examples.

Problem 1

The gas efficiency of Ariana's car is 7.8 liters per 100 kilometers. Which is the closest to the amount of liters of gas that will be needed to go a 13 kilometer distance?

(A) 9.99   **(B)** 1.00   **(C)** 1.01   **(D)** 1.02   **(E)** 1.03

Solution

Let x be the liters needed to go a 13 kilometer distance. Then we have the following proportion

$$\frac{7.8}{100} = \frac{x}{13}$$

After cross multiplying the terms we have

$$\frac{7.8}{100} = \frac{x}{13}$$
$$100x = 7.8 \cdot 13$$
$$100x = 101.4$$
$$x = 1.014$$

The closest number in the list is 1.01 and the right answer is $\boxed{\textbf{(C)} \ 1.01}$

Problem 2

Fernando is 24 centimeters taller than Lucas. How tall is Lucas if the heights of the boys satisfy the ratio 14 : 17?

(A) 110   **(B)** 112   **(C)** 114   **(D)** 116   **(E)** 118

Solution

Let x be the height of Lucas given in centimeters. Then we have the following proportion

$$\frac{x}{x+24} = \frac{14}{17}$$

After cross multiplying the terms we can solve this equation by isolating x[1]

$$\frac{x}{x+24} = \frac{14}{17}$$
$$17x = 14(x+24)$$
$$17x = 14x + 336$$
$$3x = 336$$
$$x = 112$$

and the right answer is $\boxed{(\textbf{B})\ 112}$

Problem 3

33 kilograms of sugar are needed to make $(29n-11)$ bottles of strawberry preserves, where n is some positive integer number. Which of the intervals will always contain the amount of sugar needed to make $2n-1$ bottles of preserves?

(A) $[1.5, 2.1]$ **(B)** $[1.6, 2.2]$ **(C)** $[1.7, 2.3]$ **(D)** $[1.9, 2.4]$ **(E)** $[2.0, 2.6]$

Solution

Let x be the amount of sugar needed to make $2n-1$ bottles of preserves. Then we have the following proportion

$$\frac{33}{29n-11} = \frac{x}{2n-1}$$

We can cross multiply the terms we can solve this equation for x

$$\frac{33}{29n-11} = \frac{x}{2n-1}$$
$$x \cdot (29n-11) = 33 \cdot (2n-1)$$
$$x = \frac{66n-33}{29n-11}$$

We can estimate[2] the value of x when $n=1$

$$x = \frac{66(1)-33}{29(1)-11} = \frac{11}{6} < \frac{19}{10} = 1.9$$

and, therefore, the answers **(D)** and **(E)** do not work.

---

[1] This technique is discussed in Chapter 20 "Linear Equations. Part 1"
[2] You can find more problems that involve this technique in Chapter 25 "Estimations"

We can also estimate the value of x when $n = 5$

$$x = \frac{66(5) - 33}{29(5) - 11} = \frac{297}{134} > \frac{11}{10} = 2.2$$

and, therefore, the answers **(A)** and **(B)** do not work.

This implies that the right answer is $\boxed{\textbf{(C) } [1.7, 2.3]}$

CHAPTER 4

MEAN

A **mean** (or arithmetic mean) of several numbers is defined as the sum of these numbers divided by the number of numbers. The mean value is also called the *average value*.

If we are given n numbers $a_1, a_2, \ldots, a_n$, then their mean can be found as

$$\frac{a_1 + a_2 + \ldots + a_n}{n}$$

Let us now consider several examples where we have to deal with the concept of mean.

Problem 1

Jessica took 4 tests. She scored 85, 91 and 92 on the first three tests and she forgot her score on the last test. How much did she get on the last test if the mean score of the four exams is 90?

(A) 88   (B) 89   (C) 90   (D) 91   (E) 92

Solution

Let x be the Jessica's score on the last test. Then we have the following equation

$$\frac{85 + 91 + 92 + x}{4} = 90$$

which can be solved by isolating x

$$\frac{85 + 91 + 92 + x}{4} = 90$$
$$85 + 91 + 92 + x = 360$$
$$268 + x = 360$$
$$x = 92$$

and the right answer is $\boxed{\textbf{(E) } 92}$

Problem 2

Sabrina is preparing for a math olympiad. She solved 20 problems on average per day during the first week, 18 problems per day during the next two weeks and p problems per day during the next three weeks. How many problems did she solve on average during the six weeks?

(A) $\frac{3p+56}{6}$ **(B)** $\frac{3p+49}{6}$ **(C)** $\frac{4p+56}{7}$ **(D)** $\frac{7p+42}{8}$ **(E)** $\frac{21p+390}{42}$

Solution

The number of problems solved during the first week is equal to

$$20 \cdot 7 = 140$$

The number of problems solved during the next two weeks is equal to

$$18 \cdot 14 = 252$$

The number of problems solved during the next three weeks is equal to

$$p \cdot 21 = 21p$$

Therefore, the total number of solved problems is

$$140 + 252 + 21p = 392 + 21p$$

The average number of problems per day during the six weeks is

$$\frac{392 + 21p}{42} = \frac{3p + 56}{6}$$

and the right answer is (A) $\dfrac{3p+56}{6}$

Problem 3

The mean of the numbers $a_1, a_2, \ldots, a_{99}$ is equal to 100. The mean of the numbers $a_1, a_2, \ldots, a_{99}, a_{100}$ is equal to 100.1. Find the value of a_{100}.

(A) 1   (B) 10   (C) 11   (D) 100   (E) 110

Solution

Notice that since the mean of the numbers $a_1, a_2, \ldots, a_{99}$ is equal to 100, then we have

$$\frac{a_1 + a_2 + \ldots + a_{99}}{99} = 100$$

From here

$$a_1 + a_2 + \ldots + a_{99} = 9900$$

Since the mean of the numbers $a_1, a_2, \ldots, a_{99}, a_{100}$ is equal to 100.1, then we have the following equation

$$\frac{a_1 + a_2 + \ldots + a_{99} + a_{100}}{100} = 100.1$$

which implies that

$$a_1 + a_2 + \ldots + a_{99} + a_{100} = 10010$$
$$9900 + a_{100} = 10010$$
$$a_{100} = 110$$

and the right answer is (E) 110

CHAPTER 5

MEDIAN

If there is an odd number of elements in an ordered list, then the **median** is defined as the middle number in the list. If there is an even number of elements in the list, then the **median** is defined as the mean of the two middle numbers.

For example, if $A = (1, 4, 8, 10, 11)$ and $B = (1, 4, 8, 10, 11, 19)$, then the median of the list A is 8 and the median of the list B is $\frac{8+10}{2} = 9$.

Let us now consider several examples.

Problem 1

Given a sequence of numbers

$$\frac{1}{2}, -\frac{1}{3}, \frac{1}{4}, -\frac{1}{5}, \frac{1}{6}, -\frac{1}{7}, ..., \frac{1}{40}, -\frac{1}{41}$$

Which of the following represents the median m of this sequence.

(A) 0 **(B)** $\frac{1}{40}$ **(C)** $-\frac{1}{41}$ **(D)** $\frac{1}{3280}$ **(E)** $\frac{81}{3280}$

Solution

Notice that there is a total of 40 numbers in the sequence, twenty of which are positive and the other twenty are negative. The sequence can be ordered as

$$-\frac{1}{3}, -\frac{1}{5}, -\frac{1}{7}, ..., -\frac{1}{41}, \frac{1}{40}, ..., \frac{1}{6}, \frac{1}{4}, \frac{1}{2}$$

This means that the median is the arithmetic mean of the largest negative number and the smallest positive number

$$m = \frac{-\frac{1}{41} + \frac{1}{40}}{2} = \frac{1}{3280}$$

and the right answer is $\boxed{\textbf{(D)}\ \dfrac{1}{3280}}$

Problem 2

Let A be the set of all positive integers divisible by 2 and less than 99, and B be the set of all positive integers divisible by 3 and less than 98. Jake wrote all the elements of the set A on the board and then wrote all the elements of the set B on the same board. Find the median of the numbers written on the board.

(A) 49   **(B)** 49.5   **(C)** 50   **(D)** 50.5   **(E)** 51

Solution

Let us work with the numbers less or equal to 96 apart. Let us consider the blocks of numbers of the form

$$(6k - 5, 6k - 4, 6k - 3, 6k - 2, 6k - 1, 6k)$$

where k is an integer, $1 \leq k \leq 16$.

Therefore, there are 5 numbers that will be written on the board from each block

$$(6k - 4, 6k - 3, 6k - 2, 6k, 6k)$$

Notice that there is a total of 16 blocks, resulting in 80 numbers written on the board. The last number in the last block is 96 and the only number larger than 96 written on the board is 98. Therefore, there is a total of 81 numbers written on the board and the median of this list should be the middle number. The middle number is the first number in the 9-th block, which equals $6k - 4$ for $k = 9$, i.e.

$$6(9) - 4 = 50$$

and the right answer is $\boxed{\textbf{(C)}\ 50}$

Problem 3

Given the sequence of numbers

$$1, 2, 2, 3, 3, 3, ..., 99$$

where each number is written as many times as the number represents (for example, the number 3 is written 3 times and the number 99 is written 99 times). Find the median of this sequence.

(**A**) 68   (**B**) 69   (**C**) 70   (**D**) 71   (**E**) 72

Solution

The total number of terms in this sequence can be found from the Gauss Formula as the sum of the first 99 positive integer numbers

$$1 + 2 + ... + 98 + 99 = \frac{99 \cdot 100}{2} = 4950$$

and therefore, the median is the mean of the numbers on the positions 2475 and 2476. Let us show that both terms are equal 70. Indeed, let us consider the sum of the first 70 positive integer numbers

$$1 + 2 + ... + 70 = \frac{70 \cdot 71}{2} = 2485$$

This means that the last number 70 in the sequence has the position 2485. Since the previous 69 numbers are also 70, then the numbers on the positions 2475 and 2476 are also equal 70 and the right answer is $\boxed{\textbf{(C) } 70}$

CHAPTER 6

TELESCOPING PRODUCTS

The **Telescoping Products** are the expressions, where instead of multiplying all the terms directly, we can cancel out almost all the numerators and denominators, and simplify the multiplication extensively.

Let us consider several examples to illustrate this technique.

Problem 1

Find the value of the product

$$\left(1 + \frac{1}{1}\right) \cdot \left(1 + \frac{1}{2}\right) \cdot \dots \cdot \left(1 + \frac{1}{99}\right) \cdot \left(1 + \frac{1}{100}\right)$$

(A) 50   **(B)** 100   **(C)** 101   **(D)** $\frac{1}{100}$ **(E)** $\frac{1}{101}$

Solution

Let us start by adding the fractions in each parenthesis

$$1 + \frac{1}{1} = \frac{2}{1}$$
$$1 + \frac{1}{2} = \frac{3}{2}$$
$$\cdots$$
$$1 + \frac{1}{99} = \frac{100}{99}$$
$$1 + \frac{1}{100} = \frac{101}{100}$$

Therefore, the product is equal to

$$\frac{2}{1} \cdot \frac{3}{2} \cdot \ldots \cdot \frac{100}{99} \cdot \frac{101}{100}$$

Notice that this product telescopes, i.e. all the terms in the numerator and the denominator of the fractions starting with 2 and ending with 10 can be canceled out

$$\frac{\cancel{2}}{1} \cdot \frac{\cancel{3}}{\cancel{2}} \cdot \ldots \cdot \frac{\cancel{100}}{\cancel{99}} \cdot \frac{101}{\cancel{100}} = \frac{101}{1} = 101$$

and the right answer is $\boxed{\textbf{(C) } 101}$

Problem 2

Determine the value of

$$\left(2 - \frac{2}{1000} \right) \cdot \left(2 - \frac{2}{999} \right) \cdot \ldots \cdot \left(2 - \frac{2}{3} \right) \cdot \left(2 - \frac{2}{2} \right)$$

(A) $\frac{1}{500}$ **(B)** $\frac{2^{999}}{1000}$ **(C)** $\frac{999}{1000}$ **(D)** $\frac{2^{1000}}{999}$ **(E)** $\frac{1001}{1000}$

Solution

Let us start by simplifying each parenthesis. We will factor the common factor 2 and then subtract the remaining terms

$$2 - \frac{2}{1000} = 2 \cdot \left(1 - \frac{1}{1000} \right) = 2 \cdot \frac{999}{1000}$$

$$1 - \frac{1}{999} = 2 \cdot \left(1 - \frac{1}{999}\right) = 2 \cdot \frac{998}{999}$$

$$\cdots \qquad \cdots$$

$$1 - \frac{1}{3} = 2 \cdot \left(1 - \frac{1}{3}\right) = 2 \cdot \frac{2}{3}$$

$$1 - \frac{1}{2} = 2 \cdot \left(1 - \frac{1}{2}\right) = 2 \cdot \frac{1}{2}$$

Therefore, the product is equal to

$$2^{999} \cdot \frac{999}{1000} \cdot \frac{998}{999} \cdot \ldots \cdot \frac{2}{3} \cdot \frac{1}{2}$$

Notice that this product telescopes, i.e. all the terms in the numerator and the denominator of the fractions starting with 2 and ending with 999 can be canceled out

$$2^{999} \cdot \frac{\cancel{999}}{1000} \cdot \frac{\cancel{998}}{\cancel{999}} \cdot \ldots \cdot \frac{\cancel{2}}{\cancel{3}} \cdot \frac{1}{\cancel{2}} = \frac{2^{999}}{1000}$$

and the right answer is (B) $\boxed{\dfrac{2^{999}}{1000}}$

Problem 3

Evaluate the product

$$\left(1 - \frac{1}{2^2}\right) \cdot \left(1 - \frac{1}{3^2}\right) \cdot \ldots \cdot \left(1 - \frac{1}{19^2}\right) \cdot \left(1 - \frac{1}{20^2}\right)$$

(A) $\frac{1}{20}$ (B) 1   (C) $\frac{21}{20}$ (D) $\frac{21}{40}$ (E) $\frac{1}{21}$

Solution

Notice that the general formula for each parenthesis can be rewritten as

$$1 - \frac{1}{n^2} = \frac{n^2 - 1}{n^2} = \frac{(n-1)(n+1)}{n^2} = \frac{n-1}{n} \cdot \frac{n+1}{n}$$

Let us rewrite each parenthesis as follows

$$1 - \frac{1}{2^2} = \frac{1}{2} \cdot \frac{3}{2}$$

$$1 - \frac{1}{3^2} = \frac{2}{3} \cdot \frac{4}{3}$$

$$\cdots$$

$$1 - \frac{1}{19^2} = \frac{18}{19} \cdot \frac{20}{19}$$

$$1 - \frac{1}{20^2} = \frac{19}{20} \cdot \frac{21}{20}$$

The product, therefore, is

$$\left(\frac{1}{2} \cdot \frac{3}{2} \right) \left(\frac{2}{3} \cdot \frac{4}{3} \right) \cdots \left(\frac{18}{19} \cdot \frac{20}{19} \right) \left(\frac{19}{20} \cdot \frac{21}{20} \right)$$

and can be rearranged as

$$\left(\frac{1}{2} \cdot \frac{2}{3} \cdots \cdot \frac{18}{19} \cdot \frac{19}{20} \right) \cdot \left(\frac{3}{2} \cdot \frac{4}{3} \cdots \cdot \frac{20}{19} \cdot \frac{21}{20} \right)$$

Each product telescopes and we have

$$\left(\frac{1}{\cancel{2}} \cdot \frac{\cancel{2}}{\cancel{3}} \cdots \cdot \frac{\cancel{18}}{\cancel{19}} \cdot \frac{\cancel{19}}{20} \right) \cdot \left(\frac{\cancel{3}}{2} \cdot \frac{\cancel{4}}{\cancel{3}} \cdots \cdot \frac{\cancel{20}}{\cancel{19}} \cdot \frac{21}{\cancel{20}} \right) = \frac{21}{40}$$

and the right answer is $\boxed{(\mathbf{D}) \ \dfrac{21}{40}}$

CHAPTER 7

TELESCOPING SUMS

The **Telescoping Sums** are the expressions, where instead of adding all the terms directly, we can cancel out almost all the terms and simplify the addition extensively.

Let us consider several examples to illustrate this idea.

Problem 1

Find the sum

$$\left(\frac{1}{2} - \frac{1}{3}\right) + \left(\frac{1}{3} - \frac{1}{4}\right) + \dots + \left(\frac{1}{18} - \frac{1}{19}\right) + \left(\frac{1}{19} - \frac{1}{20}\right)$$

(A) $\frac{1}{2}$ **(B)** $\frac{1}{20}$ **(C)** $\frac{1}{10}$ **(D)** $\frac{3}{20}$ **(E)** $\frac{9}{20}$

Solution

Let us group the terms in the following way

$$\frac{1}{2} + \left(-\frac{1}{3} + \frac{1}{3}\right) + \left(-\frac{1}{4} + \frac{1}{4}\right) + \dots + \left(-\frac{1}{18} + \frac{1}{18}\right) + \left(-\frac{1}{19} + \frac{1}{19}\right) - \frac{1}{20}$$

Notice that the numbers in each parenthesis cancel out and the sum telescopes

$$\frac{1}{2} + \left(-\frac{1}{\cancel{3}} + \frac{1}{\cancel{3}}\right) + \left(-\frac{1}{\cancel{4}} + \frac{1}{\cancel{4}}\right) + \dots + \left(-\frac{1}{\cancel{18}} + \frac{1}{\cancel{18}}\right) + \left(-\frac{1}{\cancel{19}} + \frac{1}{\cancel{19}}\right) - \frac{1}{20}$$

and only $\frac{1}{2}$ and $-\frac{1}{20}$ remain. Therefore, the original sum is equal to

$$\frac{1}{2} - \frac{1}{20} = \frac{9}{20}$$

and the right answer is $\boxed{\textbf{(E)} \ \dfrac{9}{20}}$

Problem 2

Determine the value of

$$\frac{1}{1 \cdot 2} + \frac{1}{2 \cdot 3} + \frac{1}{3 \cdot 4} + \dots + \frac{1}{98 \cdot 99} + \frac{1}{99 \cdot 100}$$

(A) $\frac{101}{100}$ **(B)** $\frac{1}{100}$ **(C)** $\frac{99}{100}$ **(D)** $\frac{100}{101}$ **(E)** $\frac{1}{2}$

Solution

Let us rewrite each fraction as a difference of fractions in the following way

$$\frac{1}{1 \cdot 2} = \frac{1}{1} - \frac{1}{2}$$
$$\frac{1}{2 \cdot 3} = \frac{1}{2} - \frac{1}{3}$$
$$\frac{1}{3 \cdot 4} = \frac{1}{3} - \frac{1}{4}$$
$$\dots$$
$$\frac{1}{98 \cdot 99} = \frac{1}{98} - \frac{1}{99}$$
$$\frac{1}{99 \cdot 100} = \frac{1}{99} - \frac{1}{100}$$

Therefore, the original sum is equal to

$$\frac{1}{1} - \frac{1}{2} + \frac{1}{2} - \frac{1}{3} + \frac{1}{3} - \frac{1}{4} + \dots + \frac{1}{98} - \frac{1}{99} + \frac{1}{99} - \frac{1}{100}$$

Notice that this sum telescopes

$$\frac{1}{1} - \frac{\cancel{1}}{\cancel{2}} + \frac{\cancel{1}}{\cancel{2}} - \frac{\cancel{1}}{\cancel{3}} + \frac{\cancel{1}}{\cancel{3}} - \frac{\cancel{1}}{\cancel{4}} + \cdots + \frac{\cancel{1}}{\cancel{98}} - \frac{\cancel{1}}{\cancel{99}} + \frac{\cancel{1}}{\cancel{99}} - \frac{1}{100}$$

and only $\frac{1}{1}$ and $-\frac{1}{100}$ remain. Therefore, the original sum is equal to

$$\frac{1}{1} - \frac{1}{100} = \frac{99}{100}$$

and the right answer is $\boxed{(\text{C}) \ \dfrac{99}{100}}$

Problem 3

Evaluate the sum

$$\frac{1}{3} + \frac{1}{15} + \frac{1}{35} + \frac{1}{63} + \frac{1}{99} + \frac{1}{143}$$

(A) $\frac{6}{13}$ (B) $\frac{7}{15}$ (C) $\frac{8}{17}$ (D) $\frac{5}{11}$ (E) $\frac{4}{9}$

Solution

Notice that the denominators of the fractions can be written as

$$3 = 1 \cdot 3$$
$$15 = 3 \cdot 5$$
$$35 = 5 \cdot 7$$
$$63 = 7 \cdot 9$$
$$99 = 9 \cdot 11$$
$$143 = 11 \cdot 13$$

Let us rewrite each fraction in the following way

$$\frac{1}{1 \cdot 3} = \frac{1}{2} \cdot \left(\frac{1}{1} - \frac{1}{3} \right)$$
$$\frac{1}{3 \cdot 5} = \frac{1}{2} \cdot \left(\frac{1}{3} - \frac{1}{5} \right)$$
$$\frac{1}{5 \cdot 7} = \frac{1}{2} \cdot \left(\frac{1}{5} - \frac{1}{7} \right)$$

$$\frac{1}{7 \cdot 9} = \frac{1}{2} \cdot \left(\frac{1}{7} - \frac{1}{9} \right)$$

$$\frac{1}{9 \cdot 11} = \frac{1}{2} \cdot \left(\frac{1}{9} - \frac{1}{11} \right)$$

$$\frac{1}{11 \cdot 13} = \frac{1}{2} \cdot \left(\frac{1}{11} - \frac{1}{13} \right)$$

Therefore, the original sum is equal to

$$\frac{1}{2} \cdot \left(\frac{1}{1} - \frac{1}{3} + \frac{1}{3} - \frac{1}{5} + \frac{1}{5} - \frac{1}{7} + \frac{1}{7} - \frac{1}{9} + \frac{1}{9} - \frac{1}{11} + \frac{1}{11} - \frac{1}{13} \right)$$

Notice that this sum telescopes and is equal to

$$\frac{1}{2} \cdot \left(\frac{1}{1} - \frac{\cancel{1}}{\cancel{3}} + \frac{\cancel{1}}{\cancel{3}} - \frac{\cancel{1}}{\cancel{5}} + \frac{\cancel{1}}{\cancel{5}} - \frac{\cancel{1}}{\cancel{7}} + \frac{\cancel{1}}{\cancel{7}} - \frac{\cancel{1}}{\cancel{9}} + \frac{\cancel{1}}{\cancel{9}} - \frac{\cancel{1}}{\cancel{11}} + \frac{\cancel{1}}{\cancel{11}} - \frac{1}{13} \right)$$

where only $\frac{1}{1}$ and $-\frac{1}{13}$ remain. Therefore, the original sum is equal to

$$\frac{1}{2} \cdot \left(\frac{1}{1} - \frac{1}{13} \right) = \frac{6}{13}$$

and the right answer is $\boxed{\textbf{(A)} \ \frac{6}{13}}$

CHAPTER 8

DIFFERENCE OF SQUARES

The **Difference of Squares Formula** states that any expression of the form $a^2 - b^2$ can be factored in the following way

$$a^2 - b^2 = (a - b)(a + b)$$

This formula is very useful in the problems where we are given a difference of two squares.

Let us consider several examples of its application.

Problem 1

What is the largest power of 5 that is a divisor of $2017^2 - 2012^2$?

(A) 5^0 **(B)** 5^1 **(C)** 5^2 **(D)** 5^3 **(E)** 5^4

Solution

Using the Difference of Squares Formula we have

$$2017^2 - 2012^2 = (2017 - 2012) \cdot (2017 + 2012) = 5 \cdot 4029$$

Since 4029 is not divisible by 5, then the right answer is $\boxed{\textbf{(B) } 5^1}$

Problem 2

Given that $2a - 3b = 20$ and $2a + 3b = 101$, find the value of $4a^2 - 9b^2$.

(A) 2000   (B) 2002   (C) 2020   (D) 2200   (E) 2220

Solution

Instead of finding the exact values of a and b, it is much easier to use the Difference of Squares Formula

$$4a^2 - 3b^2 = (2a - 3b) \cdot (2a + 3b) = 20 \cdot 101 = 2020$$

Thus the right answer is $\boxed{\textbf{(C) } 2020}$

Problem 3

Find the value of

$$S = (11 + 10)\left(11^2 + 10^2\right)\left(11^4 + 10^4\right)\left(11^8 + 10^8\right)$$

(A) 1   (B) 21^8 (C) 21^{16} (D) $11^{16} - 10^{16}$ (E) $11^{16} + 10^{16}$

Solution

Notice that if we multiply the initial expression by $(11 - 10)$, it will not change its value. Using the Difference of Squares Formula several times we have

$$
\begin{aligned}
S &= (11 - 10)(11 + 10)\left(11^2 + 10^2\right)\left(11^4 + 10^4\right)\left(11^8 + 10^8\right) \\
&= \left(11^2 - 10^2\right)\left(11^2 + 10^2\right)\left(11^4 + 10^4\right)\left(11^8 + 10^8\right) \\
&= \left(11^4 - 10^4\right)\left(11^4 + 10^4\right)\left(11^8 + 10^8\right) \\
&= \left(11^8 - 10^8\right)\left(11^8 + 10^8\right) \\
&= 11^{16} - 10^{16}
\end{aligned}
$$

and the right answer is $\boxed{\textbf{(D) } 11^{16} - 10^{16}}$

CHAPTER 9

SQUARE OF SUM AND SQUARE OF DIFFERENCE

The **Square of Sum Formula** states that the expression of the form $(a + b)^2$ can be expanded in the following way

$$(a + b)^2 = a^2 + 2ab + b^2$$

The **Square of Difference Formula** states that the expression of the form $(a - b)^2$ can be expanded in the following way

$$(a - b)^2 = a^2 - 2ab + b^2$$

These formulas are especially useful in the problems that include the sum or difference of the numbers, their product and the sum of their squares.

Let us consider several examples.

Problem 1

Given that $x^2 + y^2 = 7$ and $x + y = 3$. Find the value of xy.

(A) 0   **(B)** 1   **(C)** 2   **(D)** 3   **(E)** 4

Solution

Using the Square of Sum Formula we have

$$(x + y)^2 = x^2 + 2xy + y^2 = \left(x^2 + y^2\right) + 2(xy)$$

Substituting $x^2 + y^2 = 7$ and $xy = 1$ we obtain

$$(3)^2 = (7) + 2(xy)$$

From here $xy = 1$ and the right answer is $\boxed{\textbf{(B)}\ 1}$

Problem 2

The numbers b and c satisfy $b - c = 3$ and $bc = 2$. Find the value of

$$N = b^3 c + bc^3$$

(A) 9   **(B)** 13   **(C)** 18   **(D)** 24   **(E)** 26

Solution

Let us start by factoring the expression for N

$$N = b^3 c + bc^3 = bc \left(b^2 + c^2\right)$$

Notice that since we know the values of bc and $b - c$, then the value of $b^2 + c^2$ can be found from the Square of Difference Formula

$$(b - c)^2 = b^2 - 2bc + c^2$$
$$(3)^2 = b^2 - 2(2) + c^2$$
$$13 = b^2 + c^2$$

Therefore, the value of the expression N is

$$N = bc \left(b^2 + c^2\right) = 2 \cdot 13 = 26$$

and the right answer is $\boxed{\textbf{(E)}\ 26}$

Problem 3

Given that $z + \frac{1}{z} = 4$, what is the value of

$$z^5 - 14z^3 + z$$

(A) 0 **(B)** 2 **(C)** 5 **(D)** 8 **(E)** 11

Solution

Let us start by factoring z^3 from the expression

$$z^5 - 14z^3 + z = z^3 \left(z^2 - 14 + \frac{1}{z^2} \right)$$

Let us now find the value of $z^2 + \frac{1}{z^2}$ from the Square of Sum Formula

$$\left(z + \frac{1}{z} \right)^2 = z^2 + 2 + \frac{1}{z^2}$$

$$(4)^2 = z^2 + 2 + \frac{1}{z^2}$$

$$14 = z^2 + \frac{1}{z^2}$$

Therefore, the value of the expression is

$$z^3 \left(z^2 - 14 + \frac{1}{z^2} \right) = z^3 (14 - 14) = z^3 (0) = 0$$

and the right answer is $\boxed{\textbf{(A)} \, 0}$

CHAPTER 10

SUM AND DIFFERENCE OF CUBES

The **Sum of Cubes Formula** states that the expression of the form $a^3 + b^3$ can be factored in the following way

$$a^3 + b^3 = (a + b)\left(a^2 - ab + b^2\right)$$

The **Difference of Cubes Formula** states that the expression of the form $a^3 - b^3$ can be factored in the following way

$$a^3 - b^3 = (a - b)\left(a^2 + ab + b^2\right)$$

These formulas are very useful in the problems that include the sum of cubes or the difference of cubes of the numbers.

Let us consider several examples of their application.

Problem 1

Given that $x - y = \frac{3}{4}$ and $x^2 + xy + y^2 = \frac{4}{5}$. Find the value of

$$x^3 - y^3$$

(A) $\frac{1}{3}$ **(B)** $\frac{1}{5}$ **(C)** $\frac{3}{5}$ **(D)** $\frac{5}{3}$ **(E)** 1

Solution

Let us start by factoring the expression according to the Difference of Cubes Formula

$$x^3 - y^3 = (x - y)\left(x^2 + xy + y^2\right)$$

Notice that since we know $x - y$ and $x^2 + xy + y^2$, we can find the value of $x^3 - y^3$

$$x^3 - y^3 = (x - y)\left(x^2 + xy + y^2\right) = \frac{3}{4} \cdot \frac{4}{5} = \frac{3}{5}$$

and the right answer is $\boxed{\textbf{(C)}\ \dfrac{3}{5}}$

Problem 2

Given that $xy = 1$ and $x + y = 5$. Find the value of $x^3 + y^3$.

(A) 90   **(B)** 100   **(C)** 110   **(D)** 120   **(E)** 130

Solution

Using the Sum of Cubes Formula we have

$$x^3 + y^3 = (x + y)\left(x^2 - xy + y^2\right)$$

Notice that according to the Square of Sum Formula[1]

$$x^2 - xy + y^2 = x^2 + 2xy + y^2 - 3xy = (x + y)^2 - 3xy$$

Therefore, for the initial expression we have

$$x^3 + y^3 = (x + y)\left((x + y)^2 - 3xy\right) = (5)\left((5)^2 - 3\right) = 110$$

and the right answer is $\boxed{\textbf{(C)}\ 110}$ [2]

---

[1] This formula is discussed in detail in Chapter 9 "Square of Sum and Square of Difference"

[2] The same answer can be obtained in a different way applying the formula from Chapter 11 "Cube of Sum and Cube of Difference"

Problem 3

Given that $u^2 + \frac{1}{u^2} = 3$ for some positive number u. Find the value of the expression

$$u^3 + \frac{1}{u^3}$$

(A) $\sqrt{5}$ **(B)** $2\sqrt{5}$ **(C)** $8\sqrt{5}$ **(D)** 5 **(E)** $10\sqrt{2}$

Solution

Let us start with finding the value of $u + \frac{1}{u}$ from the Square of Sum Formula

$$\left(u + \frac{1}{u}\right)^2 = u^2 + 2 + \frac{1}{u^2}$$

$$\left(u + \frac{1}{u}\right)^2 = 5$$

$$u + \frac{1}{u} = \pm\sqrt{5}$$

Since u is positive, then so is $u + \frac{1}{u}$, which implies that

$$u + \frac{1}{u} = \sqrt{5}$$

Now let us factor the expression according to the Sum of Cubes Formula and substitute the values of $u + \frac{1}{u}$ and $u^2 + \frac{1}{u^2}$:

$$u^3 + \frac{1}{u^3} = \left(u + \frac{1}{u}\right)\left(u^2 - 1 + \frac{1}{u^2}\right)$$

$$= \left(\sqrt{5}\right)(3 - 1)$$

$$= 2\sqrt{5}$$

and the right answer is $\boxed{\textbf{(B) } 2\sqrt{5}}$

CHAPTER 11

CUBE OF SUM AND CUBE OF DIFFERENCE

The **Cube of Sum Formula** states that the expression of the form $(a + b)^3$ can be expanded in the following way

$$(a + b)^3 = a^3 + 3a^2b + 3ab^2 + b^3$$

The **Cube of Difference Formula** states that the expression of the form $(a - b)^3$ can be expanded in the following way

$$(a - b)^3 = a^3 - 3a^2b + 3ab^2 - b^3$$

These formulas are very useful in the problems that include the sum of cubes or the difference of cubes of the numbers.

Let us consider several examples.

Problem 1

Given an integer number n. Which of the following cannot be the value of the polynomial

$$n^3 - 6n^2 + 12n - 8$$

(A) -1 (B) 0   (C) 16   (D) 8   (E) 27

Solution

According to the Cube of Difference Formula the polynomial is equal to

$$n^3 - 6n^2 + 12n - 8 = (n)^3 - 3 \cdot (n)^2 \cdot (2) + 3 \cdot (n) \cdot (2)^2 - (2)^3 = (n - 2)^3$$

Since n is integer, then the values of the polynomial are perfect cubes. It is not hard to see that the only number in the list that is not a perfect cube is 16, and, therefore, the right answer is $\boxed{(C)\ 16}$

Problem 2

Given that $xy = 1$ and $x + y = 5$. Find the value of $x^3 + y^3$.

(A) 90   (B) 100   (C) 110   (D) 120   (E) 130

Solution

Let us apply the Cube of Sum Formula and regroup the terms in the following way

$$(x + y)^3 = x^3 + 3x^2y + 3xy^2 + y^3 = (x^3 + y^3) + 3xy(x + y)$$

Substituting the values of $x^3 + y^3$ and $x + y$ we have

$$(x + y)^3 = (x^3 + y^3) + 3 \cdot (xy)(x + y)$$
$$(5)^3 = (x^3 + y^3) + 3 \cdot (1)(5)$$
$$110 = x^3 + y^3$$

and the right answer is $\boxed{(C)\ 110}$ [1]

---

[1] The same answer can be obtained in a different way applying the formula from Chapter 10 "Sum and Difference of Cubes"

Problem 3

Given that $v - \frac{1}{v} = c$. What is the correct expression for

$$v^3 - \frac{1}{v^3}$$

(A) c^3 **(B)** $c^3 + 3c$ **(C)** $c^3 - 3c$ **(D)** $c^3 - c$ **(E)** $3c$

Solution

Let us start by applying the Cube of Difference Formula and regroup the terms in the following way

$$\left(v - \frac{1}{v}\right)^3 = v^3 - 3v + \frac{3}{v} - \frac{1}{v^3} = \left(v^3 - \frac{1}{v^3}\right) - 3\left(v - \frac{1}{v}\right)$$

Substituting the value of $v - \frac{1}{v}$ we have

$$\left(v - \frac{1}{v}\right)^3 = \left(v^3 - \frac{1}{v^3}\right) - 3\left(v - \frac{1}{v}\right)$$

$$(c)^3 = \left(v^3 - \frac{1}{v^3}\right) - 3(c)$$

$$c^3 + 3c = v^3 - \frac{1}{v^3}$$

and the right answer is $\boxed{\textbf{(B)}\ c^3 + 3c}$

CHAPTER 12

SOPHIE GERMAIN'S IDENTITY

The **Sophie Germain's Identity** states that the expression of the form $a^4 + 4b^4$ can be factored as

$$a^4 + 4b^4 = \left(a^2 + 2ab + 2b^2\right)\left(a^2 - 2ab + 2b^2\right)$$

Let us consider several examples of its application.

Problem 1

Which of the following is a factor of $81x^4 + 64y^4$ for all integers x and y?

(A) $9x^2$ **(B)** $8y^2$ **(C)** $9x^2 + 8y^2$ **(D)** $9x^2 - 4y^2$ **(E)** $9x^2 + 12xy + 8y^2$

Solution

Notice that
$$81x^4 + 64y^4 = (3x)^4 + 4(2y)^4$$

Now we can factor the expression according to Sophie Germain's Identity:

$$(3x)^4 + 4(2y)^4 = \left(9x^2 - 12xy + 8y^2\right)\left(9x^2 + 12xy + 8y^2\right)$$

Therefore, the right answer is $\boxed{\textbf{(E) } 9x^2 + 12xy + 8y^2}$

Problem 2

How many positive integer numbers x satisfy the condition that the number $x^4 + 4$ is a prime?

(A) 1   **(B)** 2   **(C)** 3   **(D)** 4   **(E)** 5

Solution

Let us start by factoring the expression according to Sophie Germain's Identity

$$x^4 + 4 = \left(x^2 - 2x + 2\right)\left(x^2 + 2x + 2\right)$$

Notice that since $x \geq 1$, then

$$x^2 + 2x + 2 \geq (1)^2 + 2 + 2 = 5$$

Therefore, the second parenthesis of the factorization cannot be equal 1 and thus the first parenthesis should equal 1, i.e.

$$x^2 - 2x + 2 = 1$$
$$x^2 - 2x + 1 = 0$$
$$(x - 1)^2 = 0$$
$$x - 1 = 0$$
$$x = 1$$

Therefore, the only such value of x is 1 and the right answer is $\boxed{\textbf{(A) } 1}$

Problem 3

Which of the following is the remainder when $1 + 2^{42}$ is divided by $2^{21} - 2^{11} + 1$?

(A) 0   **(B)** 1   **(C)** 3   **(D)** 6   **(E)** 8

Solution

Let us make a substitution $x = 2^{10}$. Therefore, we have

$$1 + 2^{42} = 1 + 4 \cdot \left(2^{10}\right)^4 = 1 + 4x^4$$

and also

$$2^{21} - 2^{11} + 1 = 2 \cdot \left(2^{10}\right)^2 - 2 \cdot \left(2^{10}\right) + 1 = 2x^2 - 2x + 1$$

Let us now find the remainder when $1 + 4x^4$ is divided by $2x^2 - 2x + 1$. According to Sophie Germain's Identity

$$1 + 4x^4 = \left(2x^2 - 2x + 1\right)\left(2x^2 + 2x + 1\right)$$

Therefore, $1 + 4x^4$ is divisible by $2x^2 - 2x + 1$, the remainder of the division is zero and the right answer is $\boxed{\textbf{(A) } 0}$

CHAPTER 13

QUADRATIC FORMULA

The **Quadratic Formula** states that the solutions of the quadratic equation

$$ax^2 + bx + c = 0 \quad (a \neq 0)$$

are given by the formula

$$x = \frac{-b \pm \sqrt{b^2 - 4ac}}{2a}$$

Let us consider several examples.

Problem 1

The solutions of the equation

$$x^2 = x + 4$$

are marked on the real number line. Find the distance between the solutions.

(A) 0   **(B)** 1   **(C)** $\sqrt{15}$ **(D)** 2   **(E)** $\sqrt{17}$

Solution

Let us start by rewriting the equation as

$$x^2 - x - 4 = 0$$

According to the Quadratic Formula the solutions of this equation are

$$x = \frac{-(-1) \pm \sqrt{(-1)^2 - 4(1)(-4)}}{2(1)} = \frac{1 \pm \sqrt{17}}{2}$$

Therefore, the distance between the solutions can be found as

$$\frac{1 + \sqrt{17}}{2} - \frac{1 - \sqrt{17}}{2} = \frac{2\sqrt{17}}{2} = \sqrt{17}$$

and the right answer is $\boxed{\textbf{(E)} \ \sqrt{17}}$

Problem 2

The solutions of the equation $3x^2 + 2x - 9 = 0$ are represented in the form $a \pm b\sqrt{7}$, where a and b are some real numbers. Which of the following is the value of $a^2 + b^2$?

(A) 5 **(B)** 9 **(C)** $\frac{1}{9}$ **(D)** $\frac{4}{9}$ **(E)** $\frac{5}{9}$

Solution

According to the Quadratic Formula the solutions of this equation are

$$x = \frac{-(2) \pm \sqrt{(2)^2 - 4(3)(-9)}}{2(3)}$$

$$= \frac{-2 \pm \sqrt{112}}{6}$$

$$= \frac{-2 \pm 4\sqrt{7}}{6}$$

$$= -\frac{1}{3} \pm \frac{2}{3} \cdot \sqrt{7}$$

Therefore, the value of $a^2 + b^2$ is equal to

$$\left(-\frac{1}{3}\right)^2 + \left(\frac{2}{3}\right)^2 = \frac{1}{9} + \frac{4}{9} = \frac{5}{9}$$

Therefore, the right answer is $\boxed{\textbf{(E)} \ \frac{5}{9}}$

Problem 3

The sum of two positive numbers is 40 and the sum of their squares 832. Which of the following is the last digit of the larger number?

(A) 6   **(B)** 4   **(C)** 2   **(D)** 7   **(E)** 5

Solution

Let the numbers be x and y, where $x > y$. Therefore, we have

$$x + y = 40$$
$$x^2 + y^2 = 832$$

Solving the first equation for y we have $y = 40 - x$. Now we substitute it into the second equation

$$x^2 + y^2 = 832$$
$$x^2 + (40 - x)^2 = 832$$
$$x^2 + 1600 - 80x + x^2 = 832$$
$$2x^2 - 80x + 768 = 0$$
$$x^2 - 40x + 384 = 0$$

According to the Quadratic Formula the solutions of this equation are

$$x = \frac{-(-40) \pm \sqrt{(-40)^2 - 4(1)(384)}}{2(1)} = \frac{40 \pm \sqrt{64}}{2} = \frac{40 \pm 8}{2}$$

The largest solution is

$$\frac{40 + 8}{2} = 24$$

and the right answer is $\boxed{\textbf{(B)} \ 4}$

CHAPTER 14

DISCRIMINANT OF A QUADRATIC EQUATION

The **Discriminant** of the quadratic equation

$$ax^2 + bx + c = 0 \quad (a \neq 0)$$

is the value defined as

$$D = b^2 - 4ac$$

The discriminant forms part of the quadratic formula

$$x = \frac{-b \pm \sqrt{b^2 - 4ac}}{2a} = \frac{-b \pm \sqrt{D}}{2a}$$

Depending on the discriminant, quadratic equation might have two real solutions, one real solution, or no real solutions

- if $D > 0$ the equation has two real solutions

- if $D = 0$ the equation has one real solution

- if $D < 0$ the equation has no real solutions

Let us consider several examples.

Problem 1

For which values of a does the equation

$$3x^2 - 5x + (a + 1) = 0$$

have only one real solution?

(A) $\frac{13}{12}$ **(B)** -1 **(C)** $\frac{7}{2}$ **(D)** 4 **(E)** 0

Solution

Notice that the equation given in the problem is quadratic with respect to x. Let us find its discriminant

$$\begin{aligned} D &= b^2 - 4ac \\ &= (-5)^2 - 4 \cdot (3) \cdot (a + 1) \\ &= 13 - 12a \end{aligned}$$

The quadratic equation has one real solution if and only if $D = 0$, i.e.

$$13 - 12a = 0$$

From here we have $a = \frac{13}{12}$ and the right answer is $\boxed{\textbf{(A)} \ \dfrac{13}{12}}$

Problem 2

How many integers n satisfy the inequality

$$n^4 + 2021 < 89n^2$$

(A) 1 **(B)** 2 **(C)** 0 **(D)** 4 **(E)** 3

Solution

Let us make a substitution $x = n^2$. Therefore, we have

$$\begin{aligned} n^4 + 2021 &< 89n^2 \\ x^2 + 2021 &< 89x \\ x^2 - 89x + 2021 &< 0 \end{aligned}$$

The left-hand side of the last inequality is a quadratic trinomial with respect to x. Let us now consider its discriminant

$$D = b^2 - 4ac$$
$$= (-89)^2 - 4 \cdot (1) \cdot (2021)$$
$$= -163$$

The discriminant is negative and, therefore, the parabola $y = x^2 - 89x + 2021$ opens upwards and does not intersect the x-axis. This means that it does not take any negative values and the right answer is $\boxed{\textbf{(C) } 0}$

Problem 3

Given the integer numbers of x and y that satisfy the equation

$$x^2 - 2x + y^2 + 2y + 1 = 0$$

Find the sum of all values that x can take.

(A) 0   **(B)** 1   **(C)** 2   **(D)** 3   **(E)** 4

Solution

Let us consider the left-hand side as a quadratic equation with respect to x

$$x^2 - 2x + \left(y^2 + 2y + 1\right) = 0$$

The quadratic equation has real roots when its discriminant is nonnegative, i.e.

$$D = b^2 - 4ac$$
$$= (-2)^2 - 4(1)\left(y^2 + 2y + 1\right)$$
$$= -4y(y + 2)$$

Since y is an integer, then last expression is nonnegative only for $y = -2, -1, 0$. It is not hard to see that

- if $y = -2$, then $x = 1$
- if $y = -1$, then $x = 0, 2$
- if $y = 0$, then $x = 1$

Therefore, the sum of the possible values of x is $0 + 1 + 2 = 3$ and the right answer is $\boxed{\textbf{(D) } 3}$ [1]

---

[1] The same answer can be obtained by completing the square. This idea is discussed in detail in Chapter 16 "Completing the Square".

CHAPTER 15

VIETA'S FORMULAS

The **Vieta's Formulas** state that if x_1 and x_2 are the roots of the quadratic equation

$$ax^2 + bx + c = 0 \quad (a \neq 0)$$

then the following equalities hold

$$x_1 + x_2 = -\frac{b}{a}$$

$$x_1 \cdot x_2 = \frac{c}{a}$$

Let us consider several examples of their application.

Problem 1

Let m and n be the solutions of the equation $5x^2 - 2x - 7 = 0$. Find the value of the expression

$$\frac{1}{m} + \frac{1}{n}$$

(A) -7 **(B)** $\frac{2}{7}$ **(C)** $-\frac{2}{7}$ **(D)** $\frac{7}{2}$ **(E)** $-\frac{7}{2}$

Solution

From the Vieta's Formulas we have

$$m + n = \frac{2}{5}$$

$$m \cdot n = -\frac{7}{5}$$

Let us now consider the expression and substitute the values of $m + n$ and $m \cdot n$

$$\frac{1}{m} + \frac{1}{n} = \frac{m + n}{mn}$$

$$= \frac{\frac{2}{5}}{-\frac{7}{5}}$$

$$= -\frac{2}{7}$$

and the right answer is $\boxed{\textbf{(E)} \; -\frac{2}{7}}$

Problem 2

Let x and y be positive real numbers that are the solutions of the quadratic equation

$$t^2 - 2xt + y = 0$$

Which of the following is the value of y?

(A) 1   **(B)** 2   **(C)** 4   **(D)** 6   **(E)** 9

Solution

From the Vieta's Formulas we have

$$x + y = 2x$$

$$x \cdot y = y$$

From the first equation we have that $x = y$. Substituting it into the second equation we have

$$x \cdot y = y$$

$$y \cdot y = y$$

$$y^2 = y$$

$$y^2 - y = 0$$

$$y(y - 1) = 0$$

Since y is positive, then this implies that $y = 1$ and the right answer is $\boxed{\textbf{(A)} \, 1}$

Problem 3

Let (u_1, u_2) and (v_1, v_2) be the points of intersection of the graphs of $y = 4x^2$ and $y = 3 - x$. Find the value of the expression

$$(u_1 + 2)(v_1 + 2)$$

(A) $\frac{3}{10}$ (B) $\frac{2}{9}$ (C) $\frac{17}{13}$ (D) $\frac{11}{4}$ (E) $\frac{13}{6}$

Solution

The x-coordinates u_1 and v_1 of the points of intersection of the given graphs satisfy the equation

$$4x^2 = 3 - x$$
$$4x^2 + x - 3 = 0$$

Now from the Vieta's Formulas we have

$$u_1 + v_1 = -\frac{1}{4}$$
$$u_1 \cdot v_1 = -\frac{3}{4}$$

Let us now expand the original expression

$$(u_1 + 2)(v_1 + 2) = u_1 v_1 + 2u_1 + 2v_1 + 4 = (u_1 v_1) + 2(u_1 + v_1) + 4$$

Therefore, substituting the values of $u_1 + v_1$ and $u_1 \cdot v_1$ we have

$$(u_1 v_1) + 2(u_1 + v_1) + 4 = \left(-\frac{3}{4}\right) + 2\left(-\frac{1}{4}\right) + 4 = \frac{11}{4}$$

and the right answer is $\boxed{\textbf{(D)} \ \frac{11}{4}}$

CHAPTER 16

COMPLETING THE SQUARE

In many problems the quadratic expression

$$ax^2 + bx + c \quad (a \neq 0)$$

should be expressed in the form

$$a(x - h)^2 + k$$

This procedure is called **Completing the Square** and relies on the Square of Sum and Square of Difference Formulas

$$(a + b)^2 = a^2 + 2ab + b^2$$
$$(a - b)^2 = a^2 - 2ab + b^2$$

Let us look at the following examples

(a) $x^2 + 6x + 9 = (x)^2 + 2(x)(3) + (3)^2 = (x + 3)^2$

(b) $x^2 - 2x + 7 = (x)^2 - 2(x)(1) + (1)^2 + 6 = (x - 1)^2 + 6$

(c) $2x^2 + 4x - 1 = 2\left((x)^2 + 2(x)(1) + (1)^2\right) - 3 = 2(x + 1)^2 - 3$

Let us now consider several problems that use this idea.

Problem 1

Find the maximum value of the expression

$$-3x^2 + 5x - 1$$

(A) $\frac{13}{12}$ (B) $-\frac{1}{2}$ (C) $\frac{5}{6}$ (D) 1 (E) 0

Solution

Let us start by completing the square

$$
\begin{aligned}
-3x^2 + 5x - 1 &= -3\left(x^2 - \frac{5}{3}x\right) - 1 \\
&= -3\left(x^2 - 2 \cdot \frac{5}{6} \cdot x + \left(\frac{5}{6}\right)^2 - \left(\frac{5}{6}\right)^2\right) - 1 \\
&= -3\left(x - \frac{5}{6}\right)^2 + 3 \cdot \left(\frac{5}{6}\right)^2 - 1 \\
&= -3\left(x - \frac{5}{6}\right)^2 + \frac{13}{12}
\end{aligned}
$$

Therefore, for all x we have[1]

$$-3\left(x - \frac{5}{6}\right)^2 + \frac{13}{12} \leq \frac{13}{12}$$

The maximum value of the expression $\frac{13}{12}$ is reached for $x = \frac{5}{6}$ and the right answer is $\boxed{\text{(A)} \ \frac{13}{12}}$

Problem 2

Given the integer numbers of x and y that satisfy the equation

$$x^2 - 2x + y^2 + 2y + 1 = 0$$

Find the sum of all values that x can take.

(A) 0 (B) 1 (C) 2 (D) 3 (E) 4

---

[1] Here we use the fact that a square of a real number is always nonnegative. This idea is discussed in detail in Chapter 24 "Obvious Inequality"

Solution

Let us start by completing the squares on the left-hand side

$$x^2 - 2x + y^2 + 2y + 1 = 0$$
$$\left(x^2 - 2x + 1\right) + \left(y^2 + 2y + 1\right) = 1$$
$$(x - 1)^2 + (y + 1)^2 = 1$$

Notice that if $(x - 1)^2 > 1$, then the last equation will not have any real solutions.

Since x and y are integers, then the last equation will have integer solutions only when $(x - 1)^2 = 1$ or $(x - 1)^2 = 0$. This implies that $x = 0$, $x = 1$ or $x = 2$. Therefore, the sum of the possible values of x is

$$0 + 1 + 2 = 3$$

and the right answer is $\boxed{\textbf{(D) } 3}$ ^{2}

Problem 3

Integer numbers a, b and c are such that the numbers $a - 2b$ and $3b - c$ are both divisible by 13. Which of the following cannot be the value of the expression

$$a^2 - 4ab + c^2 - 6bc$$

(A) 1053   **(B)** 1573   **(C)** 1872   **(D)** 1983   **(E)** 2197

Solution

Let us start by completing the squares

$$a^2 - 4ab + c^2 - 6bc = \left(a^2 - 4ab + b^2\right) + \left(c^2 - 6bc + 9b^2\right) - 13b^2$$
$$= (a - 2b)^2 + (c - 3b)^2 - 13b^2$$

Notice that since $a - 2b$ and $3b - c$ are both divisible by 13, and $13b^2$ is also divisible by 13, then the whole expression is divisible by 13. It is not hard to check that the only number in the list that is not divisible by 13 is 1983 and the right answer is $\boxed{\textbf{(D) } 1983}$

---

^{2}The same answer can be obtained in a different way by finding the discriminant of the quadratic equation. This idea is discussed in detail in Chapter 14 "Discriminant of a Quadratic Equation"

CHAPTER 17

POWERS

Expressions with **powers** are the mathematical expressions of the form

$$a^n$$

where a is usually called the base and n is called the power (or exponent).

The following properties are very useful for the problems that have expressions with powers. For all positive values of a and b, and all real values of m and n, it holds that

1. $a^m \cdot a^n = a^{m+n}$

2. $\frac{a^m}{a^n} = a^{m-n}$

3. $\left(a^m\right)^n = a^{m \cdot n}$

4. $(a \cdot b)^n = a^n \cdot b^n$

5. $\left(\frac{a}{b}\right)^n = \frac{a^n}{b^n}$

Let us now consider several examples of the problems that use powers.

Problem 1

Given the numbers $x = 16^{31}$, $y = 4^{61}$, $z = 2^{123}$. Which of the following is true?

(A) $x < y$ (B) $x < z$ (C) $y > z$ (D) $y < x$ (E) $z > x$

Solution

The number $x = 16^{31}$ can be expressed as

$$x = 16^{31} = \left(2^4\right)^{31} = 2^{4 \cdot 31} = 2^{124}$$

The number $y = 4^{61}$ can be expressed as

$$y = 4^{61} = \left(2^2\right)^{61} = 2^{2 \cdot 61} = 2^{122}$$

Therefore, we have that

$$2^{122} < 2^{123} < 2^{124}$$

This implies that

$$y < z < x$$

and the right answer is $\boxed{\textbf{(D) } y < x}$

Problem 2

Carlos wrote the number $25^{31} \cdot 32^{14}$ on the board. After this he performed all the operations and wrote the final result N. Find the sum of digits of the number N?

(A) 8   (B) 12   (C) 13   (D) 16   (E) 22

Solution

Notice that 25^{31} can be written as

$$25^{31} = \left(5^2\right)^{31} = 5^{62}$$

and 32^{14} can be written as

$$32^{14} = \left(2^5\right)^{14} = 2^{70}$$

Therefore, the number N is equal to

$$N = 25^{31} \cdot 32^{13}$$
$$= 5^{62} \cdot 2^{70}$$
$$= 5^{62} \cdot 2^{62} \cdot 2^{8}$$
$$= (5 \cdot 2)^{62} \cdot 256$$
$$= 256 \cdot 10^{62}$$

Now it is clear that the number N starts with the digits 256 followed by 62 zeros. Therefore, the sum of its digits is equal to

$$2 + 5 + 6 = 13$$

and the right answer is $\boxed{\textbf{(C) } 13}$

Problem 3

Given that

$$81^{a-b} = \frac{3^{b-c} \cdot 9^{c-a}}{27^{c}}$$

Which of the following equals to $5b - 5a$?

(A) $a - c$ **(B)** $a + 2c$ **(C)** $2c$ **(D)** $3a$ **(E)** $c - a$

Solution

Notice that

$$81^{a-b} = (3^4)^{a-b} = 3^{4a-4b}$$
$$9^{c-a} = (3^2)^{c-a} = 3^{2c-2a}$$
$$27^{c} = (3^3)^{c} = 3^{3c}$$

Therefore, the equality becomes

$$81^{a-b} = \frac{3^{b-c} \cdot 9^{c-a}}{27^{c}}$$
$$3^{4a-4b} = \frac{3^{b-c} \cdot 3^{2c-2a}}{3^{3c}}$$
$$3^{4a-4b} = \frac{3^{b+c-2a}}{3^{3c}}$$
$$3^{4a-4b} = 3^{b-2c-2a}$$

From here we have

$$4a - 4b = b - 2c - 2a$$

We can rewrite this equality as follows

$$2c = 5b - 6a$$
$$2c = 5b - 5a - a$$
$$a + 2c = 5b - 5a$$

and, therefore, the right answer is $\boxed{\textbf{(B) } a + 2c}$

CHAPTER 18

RADICALS

Expressions that represents roots of other expression are called **radicals**. For positive integer n we can define the n-th root of the number a to be the value x, such that

$$x^n = a$$

The notation for such value x is

$$x = \sqrt[n]{a}$$

The following properties are very useful for the problems that have expressions with radicals. For all real numbers $a, b \geq 0$ and positive integers m, n it holds that

1. $\sqrt[n]{a^m} = a^{\frac{m}{n}}$

2. $\sqrt[m]{\sqrt[n]{a}} = \sqrt[mn]{a}$

3. $\sqrt[n]{a \cdot b} = \sqrt[n]{a} \cdot \sqrt[n]{b}$

4. $\sqrt[n]{\frac{a}{b}} = \frac{\sqrt[n]{a}}{\sqrt[n]{b}}$ if $b \neq 0$

The Cancellation Property states that if n is odd, then

$$\sqrt[n]{a^n} = a$$

and if n is even, then

$$\sqrt[n]{a^n} = |a|$$

Let us consider several examples of the problems that use radicals.

Problem 1

Which of the following equals to

$$\sqrt{27 + 10\sqrt{2}}$$

(A) $5 + \sqrt{2}$ (B) $3 + \sqrt{2}$ (C) $1 + \sqrt{2}$ (D) $37\sqrt{2}$ (E) 39

Solution

Let us rewrite the expression under the radical and complete the square[1]

$$27 + 10\sqrt{2} = 25 + 10\sqrt{2} + 2$$
$$= (5)^2 + 2 \cdot (5) \cdot \left(\sqrt{2}\right) + \left(\sqrt{2}\right)^2$$
$$= \left(5 + \sqrt{2}\right)^2$$

Now we have

$$\sqrt{\left(5 + \sqrt{2}\right)^2} = 5 + \sqrt{2}$$

and the right answer is $\boxed{\text{(A) } 5 + \sqrt{2}}$

Problem 2

Simplify the expression

$$\frac{\sqrt[6]{8} - \sqrt[6]{9}}{\sqrt[4]{64} - \sqrt[6]{576}}$$

(A) 3 (B) 2 (C) 1 (D) $\frac{1}{2}$ (E) $\frac{1}{3}$

---

[1]The idea behind this process is discussed in detail in Chapter 16 "Completing the Square"

Solution

Notice that

$$\sqrt[6]{8} = \sqrt{\sqrt[3]{8}} = \sqrt{2}$$

$$\sqrt[6]{9} = \sqrt[3]{\sqrt{9}} = \sqrt[3]{3}$$

$$\sqrt[4]{64} = \sqrt{\sqrt{64}} = \sqrt{8} = 2\sqrt{2}$$

$$\sqrt[6]{576} = \sqrt[3]{\sqrt{576}} = \sqrt[3]{24} = 2\sqrt[3]{3}$$

Therefore, the expression becomes

$$\frac{\sqrt[6]{8} - \sqrt[6]{9}}{\sqrt[4]{64} - \sqrt[6]{576}} = \frac{\sqrt{2} - \sqrt[3]{3}}{2\sqrt{2} - 2\sqrt[3]{3}}$$

$$= \frac{\sqrt{2} - \sqrt[3]{3}}{2\left(\sqrt{2} - \sqrt[3]{3}\right)}$$

$$= \frac{\cancel{\sqrt{2} - \sqrt[3]{3}}}{2\left(\cancel{\sqrt{2} - \sqrt[3]{3}}\right)}$$

$$= \frac{1}{2}$$

and the right answer is $\boxed{\textbf{(D)}\ \dfrac{1}{2}}$

Problem 3

Find the value of x, such that

$$\sqrt[x]{\sqrt[x]{38}} = \sqrt[3x+8]{\sqrt{38}}$$

(A) 6   (B) 8   (C) 10   (D) 12   (E) 14

Solution

Notice that $\sqrt[x]{\sqrt[x]{38}}$ can be written as

$$\sqrt[x]{\sqrt[x]{38}} = \sqrt[x^2]{38}$$

and $\sqrt[3x+8]{\sqrt{38}}$ can be written as

$$\sqrt[3x+8]{\sqrt{38}} = \sqrt[6x+16]{38}$$

Therefore, we have

$$\sqrt[x^2]{38} = \sqrt[6x+16]{38}$$

which implies that

$$x^2 = 6x + 16$$
$$x^2 - 6x - 16 = 0$$
$$(x - 8)(x + 2) = 0$$

The last equality implies that $x = 8$ or $x = -2$. Since x is a positive integer, then $x = 8$ and the right answer is $\boxed{\textbf{(B) } 8}$

CHAPTER 19

ABSOLUTE VALUE

The Absolute Value of the number x is an operation defined as

$$|x| = \begin{cases} x, & \text{if } x \geq 0 \\ -x, & \text{if } x < 0 \end{cases}$$

Notice that for the nonnegative values of x, the absolute value returns the same non-negative value x. However, for the negative values of x, the absolute value returns the positive value $-x$. For example

$$|5| = 5$$
$$|-4| = 4$$
$$|0| = 0$$

The following properties are very useful for the problems that have expressions with absolute value.

For all real numbers a and b it holds that

1. $|a| \geq 0$

2. $|-a| = |a|$

3. $|a| = \sqrt{a^2}$

4. $|a|^2 = a^2$

5. $|a| = 0$ if and only if $a = 0$

6. $|a| < |b|$ if and only if $a^2 < b^2$

7. $|a \cdot b| = |a| \cdot |b|$

8. $\left|\frac{a}{b}\right| = \frac{|a|}{|b|}$ if $b \neq 0$

9. $|a + b| \leq |a| + |b|$

10. $|a - b| \geq |a| - |b|$

Let us consider several examples.

Problem 1

Which of the following is equivalent to the inequality

$$|x + 2| < |x - 2|$$

(A) $x < 0$ **(B)** $x < -1$ **(C)** $x < -2$ **(D)** $x > 0$ **(E)** $x > 2$

Solution

Let us start by squaring both sides of the given inequality

$$|x + 2| < |x - 2|$$
$$(x + 2)^2 < (x - 2)^2$$
$$x^2 + 4x + 4 < x^2 - 4x + 4$$
$$8x < 0$$
$$x < 0$$

and the right answer is $\boxed{\textbf{(A)}\ x < 0}$

Problem 2

Find the number of distinct real numbers x that satisfy the following equality

$$|x + 4| = x^2 - 5x$$

(A) 0   (B) 1   (C) 2   (D) 3   (E) 4

Solution

Case 1: if $x + 4 \geq 0$, then the absolute value should be eliminated without a change of sign

$$|x + 4| = x^2 - 5x$$
$$x + 4 = x^2 - 5x$$

which is equivalent to

$$x^2 - 6x - 4 = 0$$

We can see that this equation has two distinct real solutions that can be found from the Quadratic Formula[1]

$$x = \frac{-(-6) \pm \sqrt{(-6)^2 - 4(1)(-4)}}{2(1)}$$
$$= \frac{6 \pm \sqrt{52}}{2}$$
$$= \frac{6 \pm 2\sqrt{13}}{2}$$
$$= 3 \pm \sqrt{13}$$

Notice that both solutions satisfy the inequality $x + 4 \geq 0$.

Case 2: if $x + 4 < 0$, then the absolute value should be eliminated with a change of sign

$$|x + 4| = x^2 - 5x$$
$$-(x + 4) = x^2 - 5x$$

which is equivalent to

$$x^2 - 4x + 4 = 0$$
$$(x - 2)^2 = 0$$
$$x - 2 = 0$$
$$x = 2$$

[1] This formula is discussed in detail in Chapter 13 "Quadratic Formula"

Notice that this solution does not satisfy the inequality $x + 4 < 0$. Therefore, there are two distinct real solutions and the right answer is $\boxed{\text{(C) } 2}$

Problem 3

Find the sum of all distinct real numbers x that satisfy the equation

$$|x - 2| + |x - 1| = 2$$

(A) 0   **(B)** 3   **(C)** 5   **(D)** 7   **(E)** 9

Solution

Let us consider the cases when the expressions $x - 1$ and $x - 2$ change their sign.

Case 1: for $x \geq 2$ both absolute values are eliminated without a change of sign

$$|x - 2| + |x - 1| = 2$$
$$(x - 2) + (x - 1) = 2$$
$$2x - 3 = 2$$
$$2x = 5$$
$$x = \frac{5}{2}$$

Case 2: for $1 < x < 2$ the absolute value of $x - 1$ is eliminated without a change of sign and the absolute value of $x - 2$ is eliminated with a change of sign

$$|x - 2| + |x - 1| = 2$$
$$-(x - 2) + (x - 1) = 2$$
$$1 = 2$$

which has no solution.

Case 3: for $x < 1$ both absolute values are eliminated with a change of sign

$$|x - 2| + |x - 1| = 2$$
$$-(x - 2) - (x - 1) = 2$$
$$-2x + 3 = 2$$
$$-2x = -1$$
$$x = \frac{1}{2}$$

The sum of the distinct numbers x that satisfy the equation is equal to $\frac{5}{2} + \frac{1}{2} = 3$ and the right answer is $\boxed{\text{(B) } 3}$

CHAPTER 20

LINEAR EQUATIONS. PART 1

Linear Equations in one variable are the equations of the form

$$mx + b = c$$

where x is the unknown variable, and m, b, c are given real numbers. Linear equations are solved by *isolating* the variable x, which usually implies collecting all variables x on one side of the equation by applying the same operation to both sides of the equality.

For example, if we are given the equation

$$2(x - 1) - 3x = x - 4(x + 1)$$

we can solve it as follows

$$2(x - 1) - 3x = x - 4(x + 1)$$
$$2x - 2 - 3x = x - 4x - 4$$

$$-x - 2 = -3x - 4$$
$$-x + 3x = -4 + 2$$
$$2x = -2$$
$$x = -1$$

In many problems it is useful to call the unknown x and represent what is given in the problem in terms of a linear equation.

Let us now consider several examples.

Problem 1

Wyatt added seven consecutive integers and obtained 616. What is the smallest of these numbers?

(A) 83   **(B)** 84   **(C)** 85   **(D)** 86   **(E)** 87

Solution

Let the first integer number be x. Then the integers are

$$x, x + 1, x + 2, x + 3, x + 4, x + 5, x + 6$$

Therefore, we have the following equation

$$x + (x + 1) + \ldots + (x + 6) = 616$$

This equation can be solved by isolating x

$$7x + 21 = 616$$
$$7x = 595$$
$$x = 85$$

and the right answer is $\boxed{\textbf{(C)} \ 85}$

Problem 2

It is known that the sum of two numbers is 3 times greater than their difference. The smaller number is 1000. What is the larger number?

(A) 500   **(B)** 1000   **(C)** 1200   **(D)** 1500   **(E)** 2000

Solution

Let the larger number be x. Therefore, we have the following equation

$$x + 1000 = 3(x - 1000)$$

This equation can be solved by isolating x

$$x + 1000 = 3(x - 1000)$$
$$x + 1000 = 3x - 3000$$
$$4000 = 2x$$
$$2000 = x$$

and the right answer is $\boxed{\textbf{(E) } 2000}$

Problem 3

One side of a triangle equals half of its semiperimeter and is 2 cm smaller than the other side. Find the area of the triangle if the third side of the triangle is equal to 18.

(A) $40\sqrt{2}$ **(B)** $40\sqrt{3}$ **(C)** $50\sqrt{2}$ **(D)** $50\sqrt{3}$ **(E)** $52\sqrt{5}$

Solution

Let the first side of the triangle be x. Then the second side is $x + 2$ and the semiperimeter of the triangle is $2x$. Therefore, we have the following equation

$$\frac{1}{2}(x + x + 2 + 18) = 2x$$

This equation can be solved by isolating x

$$\frac{1}{2}(x + x + 2 + 18) = 2x$$
$$2x + 20 = 4x$$
$$20 = 2x$$
$$10 = x$$

From here the sides of the triangle are $a = 10$, $b = 12$ and $c = 18$, its semiperimeter is $p = 20$ and we can find the area of the triangle by applying the Heron's Formula[1]. The values of the expressions $p - a$, $p - b$ and $p - c$ are

$$p - a = 20 - 10 = 10$$
$$p - b = 20 - 12 = 8$$
$$p - c = 20 - 18 = 2$$

[1] This formula is discussed in detail in Chapter 55 "Heron's Formula"

By Heron's Formula the area of the triangle is equal to

$$S = \sqrt{20 \cdot 10 \cdot 8 \cdot 2} = \sqrt{3200} = 40\sqrt{2}$$

and the right answer is $\boxed{\textbf{(A)} \; 40\sqrt{2}}$

CHAPTER 21

LINEAR EQUATIONS. PART 2

Linear Equations in one variable are the equations of the form

$$mx + b = c$$

where x is the unknown variable, and m, b, c are assumed to be given. Linear equations are solved by *isolating* the variable x, which usually implies applying the same operation to both sides of the equality.

In many problems the given equation might be equivalent to some linear equation in terms of one of the variables. Therefore, the natural approach would be to start the solution of the problem by solving the linear equation.

Let us consider several examples of how this technique can be applied.

Problem 1

Find the sum of all values of the parameter a, such that the equation

$$(x - a)(x + 2) = (x - 3)(x + 3)$$

has no solution for x.

(**A**) -8 (**B**) -3 (**C**) 0 (**D**) 1 (**E**) 2

Solution

Notice that after the distribution the equation will be linear in terms of x and we will be able to solve it as follows

$$(x - a)(x + 2) = (x - 3)(x + 3)$$
$$x^2 - ax + 2x - 2a = x^2 + 3x - 3x - 9$$
$$-ax + 2x = 2a - 9$$
$$x(-a + 2) = 2a - 9$$

If $a = 2$, then the equation becomes $0 = -5$ and has no solutions. If $a \neq 2$, then the equation has solution

$$x = \frac{2a - 9}{2 - a}$$

Therefore, the initial equation has no solution for x only for $a = 2$ and the right answer is $\boxed{\textbf{(E) } 2}$

Problem 2

Given that $\frac{2x-y}{2x+y} = 3$. Find the value of

$$\frac{3x - y}{3x + y}$$

(**A**) 2 (**B**) 4 (**C**) 6 (**D**) 8 (**E**) 10

Solution

Start by noticing that the condition

$$\frac{2x - y}{2x + y} = 3$$

represents a linear equation in terms of y. We will now solve this equation for y

$$\frac{2x - y}{2x + y} = 3$$

$$2x - y = 3(2x + y)$$

$$2x - y = 6x + 3y$$

$$-y - 3y = 6x - 2x$$

$$-4y = 4x$$

$$y = -x$$

Now let us substitute $y = -x$ into the expression given in the problem

$$\frac{3x - y}{3x + y} = \frac{3x - (-x)}{3x + (-x)}$$

$$= \frac{4x}{2x}$$

$$= 2$$

and the right answer is $\boxed{\textbf{(A)}\ 2}$

Problem 3

How many pairs (m, n) of integer numbers satisfy the equality

$$m^2 - mn = 5m - n$$

(A) 0   **(B)** 2   **(C)** 4   **(D)** 6   **(E)** 8

Solution

Notice that the given equation is linear in terms of n. Let us now solve it for n

$$m^2 - mn = 5m - n$$

$$-mn + n = 5m - m^2$$

$$n(-m + 1) = 5m - m^2$$

$$n = \frac{5m - m^2}{-m + 1}$$

$$n = \frac{m^2 - 5m}{m - 1}$$

We can now represent this expression as

$$\frac{m^2 - 5m}{m - 1} = \frac{(m - 1)(m - 4) - 4}{m - 1} = m - 4 - \frac{4}{m - 1}$$

The last expression is integer if and only if $m - 1$ divides 4. The possible integer divisors of 4 are ± 1, ± 2 and ± 4. This gives a total of 6 integer values of m, each having its corresponding value of n. Therefore, the right answer is $\boxed{\textbf{(D)}\ 6}$

CHAPTER 22

SYSTEMS OF EQUATIONS. PART 1

Systems of Equations represent a set of two or more equations considered together. Usually the systems of equations have more than one variable. The two most common ways to solve a system of equations are the *method of combination* and the *method of substitution*.

For example, let us assume that we are given the following system of equations

$$\begin{cases} x + y = 3 \\ x - y = 1 \end{cases}$$

The *method of combination* states that we can *combine* these equations, for example, by adding them and obtain only one equation in one variable

$$(x + y) + (x - y) = (3) + (1)$$
$$2x = 4$$
$$x = 2$$

Then we can find the value of y from the first equation

$$x + y = 3$$
$$(2) + y = 3$$
$$y = 1$$

Alternatively we can apply the *method of substitution* by isolating x in the second equation

$$x - y = 1$$
$$x = y + 1$$

and then *substitute* it into the first equation to find the value of y

$$x + y = 3$$
$$(y + 1) + y = 3$$
$$2y = 2$$
$$y = 1$$

Then we can find the value of x from the first equation

$$x + y = 3$$
$$x + (1) = 3$$
$$x = 2$$

In many problems it is useful to call the unknowns x and y and represent what is given in the problem in terms two or more equations.

Let us consider several examples.

Problem 1

The sum of two numbers is equal 111. Find the larger of these numbers if their difference is 33.

(A) 36   (B) 72   (C) 78   (D) 99   (E) 100

Solution

Let the larger number be x and the smaller number be y. Therefore, we have the following system of equations

$$\begin{cases} x + y = 111 \\ x - y = 33 \end{cases}$$

By adding these equations we have

$$(x + y) + (x - y) = 111 + 33$$
$$2x = 144$$
$$x = 72$$

and the right answer is $\boxed{\textbf{(B) } 72}$

Problem 2

Andrea has a farm, where she has chickens and pigs. One day she counted 100 heads and 320 legs. How many chickens are on the farm?

(A) 20   **(B)** 40   **(C)** 60   **(D)** 80   **(E)** 100

Solution

Let x be the number of chickens and y be the number of pigs. Therefore, we have the following system of equations

$$\begin{cases} x + y = 100 \\ 2x + 4y = 320 \end{cases}$$

From the first equation we have

$$x + y = 100$$
$$y = 100 - x$$

Now we substitute $y = 100 - x$ into the second equation

$$2x + 4y = 320$$
$$2x + 4(100 - x) = 320$$

This equation can be solved by isolating x

$$2x + 4(100 - x) = 320$$
$$2x + 400 - 4x = 320$$
$$-2x = -80$$
$$x = 40$$

and the right answer is $\boxed{\textbf{(B) } 40}$

Problem 3

There are 72 pencils. Some are green, some are blue and some are yellow. It is known that if we double the number of the blue pencils, then it will still be 3 short of the number of green pencils. Find the number of blue pencils if there are 12 yellow pencils in the box.

(A) 12   (B) 15   (C) 18   (D) 19   (E) 20

Solution

Let x be the number of blue pencils and y be the number of green pencils. Therefore, we have the following system of equations

$$\begin{cases} x + y + 12 = 72 \\ 2x + 3 = y \end{cases}$$

Notice that the expression for y given in the second equation as $2x + 3 = y$ can be substituted into the first equation

$$x + y + 12 = 72$$
$$x + (2x + 3) + 12 = 72$$

This equation can be solved by isolating x

$$x + 2x + 3 + 12 = 72$$
$$3x + 15 = 72$$
$$3x = 57$$
$$x = 19$$

and the right answer is $\boxed{\textbf{(D) } 19}$

CHAPTER 23

SYSTEMS OF EQUATIONS. PART 2

In many problems we are given several equations considered together. We will say that these equations form a **System of Equations**. Instead of solving the system, sometimes, it is useful to combine the equations and obtain new potentially useful equalities.

Let us consider several examples.

Problem 1

The numbers a, b and c satisfy the system of equations

$$\begin{cases} a + b - c = 2019 \\ b + c - a = 2020 \\ c + a - b = 2021 \end{cases}$$

Find the value of $a + b + c$.

(**A**) 5880　　(**B**) 5940　　(**C**) 6000　　(**D**) 6060　　(**E**) 6120

Solution

Instead of trying to solve the system, let us add the three equations

$$(a + b - c) + (b + c - a) + (c + a - b) = 2019 + 2020 + 2021$$
$$a + b + c = 6060$$

and the right answer is $\boxed{\textbf{(D)}\ 6060}$

Problem 2

If the numbers x, y and z satisfy the system of equations

$$\begin{cases} 2x + 3y = 4z + 1 \\ 2x - 3y = 4z - 1 \end{cases}$$

then the value of $4x^2 + 9y^2 - 1$ is equal to

(A) $16z^2$ **(B)** $16z^2 - 1$ **(C)** $16z^2 + 1$ **(D)** $32z^2$ **(E)** $32z^2 + 2$

Solution

Let us start by squaring the first equation

$$2x + 3y = 4z + 1$$
$$(2x + 3y)^2 = (4z + 1)^2$$
$$4x^2 + 12xy + 9y^2 = 16z^2 + 8z + 1$$

Let us now square the second equation

$$2x - 3y = 4z - 1$$
$$(2x - 3y)^2 = (4z - 1)^2$$
$$4x^2 - 12xy + 9y^2 = 16z^2 - 8z + 1$$

By adding the results we have

$$\left(4x^2 + 12xy + 9y^2\right) + \left(4x^2 - 12xy + 9y^2\right) = \left(16z^2 + 8z + 1\right) + \left(16z^2 + 8z + 1\right)$$
$$8x^2 + 18y^2 = 32z^2 + 2$$
$$4x^2 + 9y^2 = 16z^2 + 1$$
$$4x^2 + 9y^2 - 1 = 16z^2$$

and the right answer is $\boxed{\textbf{(A)}\ 16z^2}$

Problem 3

The numbers u, v and w satisfy the equations

$$\begin{cases} u^2 + uv = v + w \\ v^2 + vw = w + u \\ w^2 + wu = u + v \end{cases}$$

How many distinct values can the product uvw take?

(A) 1   **(B)** 2   **(C)** 4   **(D)** 8   **(E)** 16

Solution

Let us start by factoring the left-hand side of each equation

$$u(u + v) = v + w$$
$$v(v + w) = w + u$$
$$w(w + u) = u + v$$

Now let us multiply the three equations

$$uvw(u + v)(v + w)(w + u) = (u + v)(v + w)(w + u)$$

If none of the expressions $u + v$, $v + w$, $w + u$ is equal to zero, then this equation is equivalent to

$$uvw = 1$$

Let us now assume that, for example, $u + v = 0$. This implies that from the first equation it should hold that $v + w = 0$. This in turn implies that from the second equation it should hold that $w + u = 0$. Therefore, we have a system of equations

$$\begin{cases} u + v = 0 \\ v + w = 0 \\ w + u = 0 \end{cases}$$

By adding these three equations we have $u + v + w = 0$. From here we obtain $u = v = w = 0$ and, therefore, we have

$$uvw = 0$$

Therefore, the product uvw can only take two distinct values and the right answer is

$\boxed{\textbf{(B)}\ 2}$

CHAPTER 24

OBVIOUS INEQUALITY

Given a real number x, the inequality of the form

$$x^2 \geq 0$$

is known as the **Obvious Inequality**. It comes from the fact that the square of any real number is always a nonnegative number. In general, x can denote any expression containing numbers, variables and operations. Many problems are based on the procedure of completing the square and then use the obvious inequality.

Let us now consider several examples.

Problem 1

Find the minimum value of the expression

$$a^2 - 6a + b^2 - 18b$$

(A) -180 **(B)** -120 **(C)** -90 **(D)** 0 **(E)** 30

Solution

Let us start by completing the square[1]

$$a^2 - 6a + b^2 - 18b = \left(a^2 - 6a + 9\right) + \left(b^2 - 18b + 81\right) - 90$$
$$= (a - 3)^2 + (b - 9)^2 - 90$$

Since $(a - 3)^2 \geq 0$ and $(b - 9)^2 \geq 0$, then

$$(a - 3)^2 + (b - 9)^2 - 90 \geq -90$$

The value -90 is reached for $a = 3$ and $b = 9$. Therefore, the right answer is
$\boxed{\textbf{(C)} -90}$

Problem 2

Which of the following expressions is less or equal to $x^2 y^2$ for all real values of the variables x and y?

(A) $2xy$ **(B)** $-x - y$ **(C)** 1 **(D)** $\frac{x}{y}$ **(E)** $2xy - 1$

Solution

The inequality $2xy \leq x^2 y^2$ becomes false for $x = 1$, $y = 1$, because

$$2 > 1$$

The inequality $-x - y \leq x^2 y^2$ becomes false for $x = -\frac{1}{2}$, $y = -\frac{1}{2}$, because

$$1 > \frac{1}{16}$$

The inequality $1 \leq x^2 y^2$ becomes false for $x = 0$, $y = 0$, because

$$1 > 0$$

The inequality $\frac{x}{y} \leq x^2 y^2$ becomes false for $x = \frac{1}{2}$, $y = \frac{1}{2}$, because

$$1 > \frac{1}{16}$$

The inequality $2xy - 1 \leq x^2 y^2$ can be rewritten as

$$2xy - 1 \leq x^2 y^2$$
$$0 \leq x^2 y^2 - 2xy + 1$$
$$0 \leq (xy - 1)^2$$

---

[1]This technique is discussed in detail in Chapter 16 "Completing the Square"

The last inequality represent the Obvious Inequality and, therefore, the right answer is $\boxed{\textbf{(E) } 2xy - 1}$

Problem 3

Positive real numbers x, y and z satisfy the equation
$$(xy - \sqrt{10})^2 + (yz - \sqrt{15})^2 + (zx - \sqrt{24})^2 = 0$$
Which of the following is the value of xyz?

(A) 30   **(B)** 48   **(C)** 56   **(D)** 60   **(E)** 72

Solution

Notice that by the Obvious Inequality inequality
$$(xy - \sqrt{10})^2 \geq 0$$
$$(yz - \sqrt{15})^2 \geq 0$$
$$(zx - \sqrt{24})^2 \geq 0$$
and, therefore, the left-hand side of the equation is always nonnegative. The expression is equal to zero if and only if each expression under the square is zero, i.e.
$$xy - \sqrt{10} = 0$$
$$yz - \sqrt{15} = 0$$
$$zx - \sqrt{24} = 0$$
This represents a system of equations[2] that can be rewritten as
$$\begin{cases} xy = \sqrt{10} \\ yz = \sqrt{15} \\ zx = \sqrt{24} \end{cases}$$
By multiplying these three equations we have
$$xy \cdot yz \cdot zx = \sqrt{10} \cdot \sqrt{15} \cdot \sqrt{24}$$
$$(xyz)^2 = \sqrt{3600}$$
$$xyz = \pm 60$$

Since x, y and z are positive, then $xyz = 60$ and the right answer is $\boxed{\textbf{(D) } 60}$

---

[2] You can find more problems that involve these types of systems of equations in Chapter 23 "Systems of Equations. Part 2"

CHAPTER 25

ESTIMATIONS

In many problems we are required to make **estimations** in order to deduce which of the given numbers is larger than the other. These estimations usually take the form of inequalities.

In practice it is good to remember that if a, b, c and d are real numbers, such that $a > b$ and $d > 0$, then the following inequalities hold

1. $a + c > b + c$

2. $a - c > b - c$

3. $a \cdot d > b \cdot d$

4. $\frac{a}{d} > \frac{b}{d}$

Let us consider several examples.

Problem 1

Which of the following numbers is not larger than 0.4?

(A) $\left(\frac{15}{7}\right)^{-1}$ **(B)** $\frac{\sqrt{10}}{8}$ **(C)** $\frac{\pi-1}{5}$ **(D)** $\frac{41}{100}$ **(E)** 0.7^2

Solution

Notice that $\frac{41}{100}$ is larger than 0.4. Indeed

$$\frac{41}{100} > \frac{40}{100} = 0.4$$

Also 0.7^2 is larger than 0.4 since

$$0.7^2 = 0.49 > 0.4$$

Since the power -1 represents a reciprocal of a fraction, then

$$\left(\frac{15}{7}\right)^{-1} = \frac{7}{15}$$

which is also greater than 0.4, since

$$\frac{7}{15} > \frac{6}{15} = \frac{2}{5} = 0.4$$

Notice that $\frac{\pi-1}{5}$ is also greater than 0.4

$$\frac{\pi-1}{5} > \frac{3-1}{5} > \frac{2}{5} = 0.4$$

Finally, let us confirm that the number $\frac{\sqrt{10}}{8}$ is smaller than 0.4. Indeed, the inequality $\frac{\sqrt{10}}{8} < 0.4$ is equivalent to the following series of inequalities

$$\frac{\sqrt{10}}{8} < \frac{2}{5}$$
$$\frac{10}{64} < \frac{4}{25}$$
$$10 \cdot 25 < 4 \cdot 64$$
$$250 < 256$$

which is obviously true. Therefore, the right answer is $\boxed{\textbf{(B) } \dfrac{\sqrt{10}}{8}}$

Problem 2

Which of the following is true for the numbers $a = 0.00001^{\frac{2}{5}}$, $b = 0.1^{\sqrt{2}}$ and $c = 0.01^{0.9}$?

(A) $c > a$ **(B)** $a < b$ **(C)** $c > b$ **(D)** $c < c$ **(E)** $a < c$

Solution

Let us start by rewriting the number a with base 0.1

$$x = 0.00001^{\frac{2}{5}}$$
$$= \left((0.1)^4\right)^{\frac{2}{5}}$$
$$= 0.1^{\frac{8}{5}}$$

and the number c with base 0.1

$$x = 0.01^{0.9}$$
$$= \left((0.1)^2\right)^{0.9}$$
$$= 0.1^{1.8}$$

Now we just need to compare the numbers $x = \frac{8}{5}$, $y = \sqrt{2}$ and $z = 1.8$.

It is easy to compare the numbers x and z

$$x = \frac{8}{5} = \frac{16}{10} = 1.6 < 1.8 = z$$

and, therefore, $x < z$.

Let us now compare the numbers x and y and show that $x > y$. Indeed, the inequality $x > y$ is equivalent to the following series of inequalities

$$x > y$$
$$\frac{8}{5} > \sqrt{2}$$
$$\left(\frac{8}{5}\right)^2 > 2$$
$$\frac{64}{25} > 2$$
$$64 > 50$$

which is obviously true.

Therefore, $x > y$ and we have

$$a = 0.1^x < 0.1^y = b$$

and the right answer is $\boxed{\textbf{(B) } a < b}$

Problem 3

Which of the following numbers is the largest?

(A) 2^{48} **(B)** 4^{23} **(C)** 3^{32} **(D)** 8^{15} **(E)** 10^{17}

Solution

Notice that it is easy to compare the numbers 2^{48}, 4^{23} and 8^{15}. Indeed

$$8^{15} = \left(2^3\right)^{15} = 2^{45}$$

and

$$4^{23} = \left(2^2\right)^{23} = 2^{46}$$

Since

$$2^{45} < 2^{46} < 2^{48}$$

then we have

$$8^{15} < 4^{23} < 2^{48}$$

Let us now compare the numbers 3^{32} and 10^{17}

$$3^{32} = 3^{2\cdot 16} = \left(3^2\right)^{16} = 9^{16} < 10^{16} < 10^{17}$$

which implies that $3^{32} < 10^{17}$.

Finally, let us compare the numbers 10^{17} and 2^{48}

$$2^{48} = 2^{3\cdot 16} = \left(2^3\right)^{16} = 8^{16} < 10^{16} < 10^{17}$$

which implies that $2^{48} < 10^{17}$.

Therefore, the largest number is 10^{17} and the right answer is $\boxed{\textbf{(E) } 10^{17}}$

CHAPTER 26

SEQUENCES

Sequence is an ordered list of objects called *terms*. Each sequence has a name, usually denoted by a letter of some alphabet, and a nonnegative number called *index* representing the position of the term in the sequence. For example

$$a_1 - 1^{\text{st}} \text{ element}$$

$$a_2 - 2^{\text{nd}} \text{ element}$$

$$\cdots$$

$$a_n - n^{\text{th}} \text{ element}$$

Sequences may be finite and infinite. In either case a sequence can be defined using the explicit formula of the n-th term. For example

$$a_n = 3n + 4$$

A sequence can also be defined by giving its first term and providing a *recurrence relation* to construct each element in terms of the ones before it. For example

$$a_1 = 7$$

$$a_{n+1} = a_n + 3$$

Notice that both examples define the same sequence

$$7, 10, 13, 16, 19, ...$$

Let us consider several examples.

Problem 1

The sequence x_n is the given by the following rule

$$x_{n+1} = x_n + n$$

Find the value of x_{40} if $x_1 = 10$.

(A) 780   (B) 790   (C) 800   (D) 810   (E) 820

Solution

Notice that we can find the terms of the sequence from its recurrence relation

$$x_2 = x_1 + 1 = 10 + 1$$
$$x_3 = x_2 + 2 = 10 + 1 + 2$$
$$x_4 = x_3 + 3 = 10 + 1 + 2 + 3$$
$$...$$
$$x_{39} = x_{38} + 38 = 10 + 1 + 2 + 3 + ... + 38$$
$$x_{40} = x_{39} + 39 = 10 + 1 + 2 + 3 + ... + 38 + 39$$

By applyng Gauss Formula we have

$$x_{40} = 10 + (1 + 2 + 3 + ... + 38 + 39) = 10 + \frac{39 \cdot 40}{2} = 790$$

and the right answer is $\boxed{\textbf{(B) } 790}$

Problem 2

The sequence a_n is defined as

$$a_n = \frac{5n + 5}{n}$$

Find the product of the first 10 terms of this sequence.

(A) 11   (B) 5^{10} (C) 5^{11} (D) $11 \cdot 5^{10}$ (E) $11 \cdot 5^{11}$

Solution

Notice that we can find the terms of the sequence from its recurrence relation

$$a_1 = \frac{5(1) + 5}{1} = \frac{10}{1} = 5 \cdot \frac{2}{1}$$

$$a_2 = \frac{5(2) + 5}{2} = \frac{15}{2} = 5 \cdot \frac{3}{2}$$

...

$$a_9 = \frac{5(9) + 5}{9} = \frac{50}{9} = 5 \cdot \frac{10}{9}$$

$$a_{10} = \frac{5(10) + 5}{10} = \frac{55}{10} = 5 \cdot \frac{11}{10}$$

Therefore, the product of the first 10 terms of this sequence is equal to

$$\left(5 \cdot \frac{2}{1}\right) \left(5 \cdot \frac{3}{2}\right) \cdots \left(5 \cdot \frac{10}{9}\right) \left(5 \cdot \frac{11}{10}\right)$$

which can be rearranged and telescoped[1]

$$5^{10} \cdot \left(\frac{2}{1} \cdot \frac{3}{2} \cdot \ldots \cdot \frac{10}{9} \cdot \frac{11}{10}\right) = 5^{10} \cdot \left(\frac{\cancel{2}}{1} \cdot \frac{\cancel{3}}{\cancel{2}} \cdot \ldots \cdot \frac{\cancel{10}}{\cancel{9}} \cdot \frac{11}{\cancel{10}}\right)$$

and the right answer is $\boxed{\textbf{(D)} \ 11 \cdot 5^{10}}$

Problem 3

The sequence u_n is defined as $u_1 = 1$ and for all positive integer n

$$u_{n+1} = u_n + \frac{1}{u_n}$$

What is the smallest integer value of m, such that $u_m > 3.5$?

(A) 4   **(B)** 5   **(C)** 6   **(D)** 7   **(E)** 8

Solution

It is not hard to see that the first 4 terms of the sequence are all less than 3.5. Indeed

$$u_1 = 1 < 4$$

$$u_2 = u_1 + \frac{1}{u_1} = 1 + 1 = 2 < 3.5$$

---

[1] This technique is discussed in detail in Chapter 6 "Telescoping Products"

$$u_3 = u_2 + \frac{1}{u_2} = 2 + \frac{1}{2} = \frac{5}{2} = 2.5 < 3.5$$

$$u_4 = u_3 + \frac{1}{u_3} = \frac{5}{2} + \frac{2}{5} = \frac{29}{10} = 2.9 < 3.5$$

Let us use estimations[2] to show that $u_5 < 3.5$. Indeed

$$
\begin{aligned}
u_5 &= u_4 + \frac{1}{u_4} \\
&= \frac{29}{10} + \frac{10}{29} \\
&= \frac{941}{290} \\
&< \frac{1015}{290} \\
&= 3.5
\end{aligned}
$$

Let us use estimations to show that $u_6 > 3.5$. Indeed

$$
\begin{aligned}
u_6 &= u_5 + \frac{1}{u_5} \\
&= \frac{941}{290} + \frac{290}{941} \\
&= 3 + \frac{71}{290} + \frac{290}{941} \\
&> 3 + \frac{71}{290} + \frac{290}{941} \\
&> 3 + 0.21 + 0.29 \\
&= 3.5
\end{aligned}
$$

The last inequality follows from

$$\frac{71}{290} > \frac{63}{300} = \frac{21}{100} = 0.21$$

and

$$\frac{290}{941} > \frac{290}{940} = \frac{29}{94} > \frac{29}{100} = 0.29$$

Therefore, $u_6 > 3.5$ and the right answer is $\boxed{\text{(C) } 6}$

---

[2]This technique is discussed in detail in Chapter 25 "Estimations"

CHAPTER 27

ARITHMETIC SEQUENCE

Arithmetic Sequence is a sequence defined by the recurrence relation

$$a_1 = a$$
$$a_n = a_{n-1} + d$$

where a and d are some given real numbers. The number a is called the *initial term* and the number d is called the *common difference*.

The formula for the n-th term a_n of an arithmetic sequence is

$$a_n = a_1 + (n-1)d$$

Note that from here we immediately have that every the n-th term a_n is the arithmetic mean of its neighboring terms a_{n-1} and a_{n+1}

$$a_n = \frac{a_{n-1} + a_{n+1}}{2}$$

The formula for the sum S_n of the first n terms of an arithmetic sequence is

$$S_n = \left(\frac{a_1 + a_n}{2}\right) \cdot n$$

Let us consider several examples.

Problem 1

Find the number of terms in the arithmetic sequence

$$8, 20, 32, \ldots, 584, 596, 608$$

(A) 48   **(B)** 49   **(C)** 50   **(D)** 51   **(E)** 52

Solution

Notice that in this problem the first term is $a_1 = 8$, the n-th term is $a_n = 608$. The common difference can be found as

$$d = a_2 - a_1 = 20 - 8 = 12$$

Therefore, from the formula for the n-th term a_n of an arithmetic sequence we have

$$a_n = a_1 + (n-1)d$$
$$608 = 8 + (n-1) \cdot 12$$
$$600 = (n-1) \cdot 12$$
$$50 = n - 1$$
$$51 = n$$

and the right answer is $\boxed{\textbf{(D)} \ 51}$

Problem 2

The terms of an arithmetic sequence satisfy the equalities

$$a_{10} + \ldots + a_{19} = 19$$
$$a_{20} + \ldots + a_{29} = 29$$

What is the common difference of this arithmetic sequence?

(A) 0.1   **(B)** 0.5   **(C)** 1   **(D)** 1.1   **(E)** 10

Solution

Let the first term of the sequence be a and the common difference be d. Notice that from the formula for the n-th term of an arithmetic sequence we can find the difference terms a_m and a_n

$$a_m = a + (m-1)d$$
$$a_n = a + (n-1)d$$

and therefore

$$a_m - a_n = (m - n)d$$

From here we have

$$a_{20} - a_{10} = (20 - 10)d = 10d$$
$$a_{21} - a_{11} = (21 - 11)d = 10d$$
$$...$$
$$a_{28} - a_{18} = (28 - 18)d = 10d$$
$$a_{29} - a_{19} = (29 - 19)d = 10d$$

Therefore, subtracting the two equations given in the problem we have

$$(a_{20} + ... + a_{29}) - (a_{10} + ... + a_{19}) = 29 - 19$$
$$(a_{20} - a_{10}) + ... + (a_{29} - a_{19}) = 10$$
$$(10d) + ... + (10d) = 10$$
$$100d = 10$$
$$d = 0.1$$

and the right answer is $\boxed{\textbf{(A)}\ 0.1}$

Problem 3

The first three terms of an arithmetic sequence are

$$k^2, 4k + 6, k^2 + 20$$

What is the sum of the first 10 terms of the sequence?

(A) 94   **(B)** 214   **(C)** 490   **(D)** 510   **(E)** 640

Solution

Since the n-th term a_n is the arithmetic mean of the neighbors a_{n-1} and a_{n+1}, the we have

$$a_2 = \frac{a_1 + a_3}{2}$$
$$4k + 6 = \frac{(k^2) + (k^2 + 20)}{2}$$
$$8k + 12 = 2k^2 + 20$$
$$0 = 2k^2 - 8k + 8$$

$$0 = k^2 - 4k + 4$$
$$0 = (k - 2)^2$$
$$0 = k - 2$$
$$2 = k$$

From here we have

$$a_1 = k^2 = (2)^2 = 4$$
$$a_2 = 4k + 6 = 4(2) + 6 = 14$$

and the common difference is $d = a_2 - a_1 = 10$.

We can now find the term a_{10} as follows

$$a_{10} = 4 + (10 - 1) \cdot 10 = 94$$

Therefore, from the formula for sum of the first n terms of an arithmetic sequence, we have

$$S_{10} = \left(\frac{a_1 + a_{10}}{2} \right) \cdot 10$$
$$= \left(\frac{4 + 94}{2} \right) \cdot 10$$
$$= 490$$

and the right answer is $\boxed{\textbf{(C)} \ 490}$

CHAPTER 28

GEOMETRIC SEQUENCE

Geometric Sequence is a sequence defined by the recurrence relation

$$g_1 = g$$
$$g_n = g_{n-1} \cdot r$$

where g and r are some given nonzero real numbers. The number g is called the *initial term* and the number r is called the *common ratio*.

The formula for the n-th term g_n of an arithmetic sequence is

$$g_n = g_1 \cdot r^{n-1}$$

Note that from here we immediately have that every the n-th term g_n is the geometric mean of its neighboring terms g_{n-1} and g_{n+1}

$$g_n = \sqrt{g_{n-1} \cdot g_{n+1}}$$

The formula for the sum S_n of the first n terms of an geometric sequence is

$$S_n = \left(\frac{1 - r^n}{1 - r} \right) \cdot g_1$$

The formula for the sum of an infinite geometric series is

$$S = \frac{g_1}{1 - r}$$

Let us now consider several examples.

Problem 1

The terms of the geometric sequence x_n satisfy the following condition

$$x_{11} \cdot x_{21} = 2x_{31}$$

Find the first term of the sequence.

(A) 1   (B) 2   (C) 4   (D) 8   (E) $\frac{1}{2}$

Solution

Let x_1 be the first term and the r be the common ratio of the sequence. Then from the formula for the n-th term of a geometric sequence we have

$$x_{11} = x_1 \cdot r^{11-1} = x_1 r^{10}$$
$$x_{21} = x_1 \cdot r^{21-1} = x_1 r^{20}$$
$$x_{31} = x_1 \cdot r^{31-1} = x_1 r^{30}$$

Therefore, the equality given in the problem can be rewritten as

$$x_1 r^{10} \cdot x_1 r^{20} = 2x_1 r^{30}$$
$$x_1^2 r^{30} = 2x_1 r^{30}$$
$$x_1^2 r^{30} - 2x_1 r^{30} = 0$$
$$x_1 r^{30}(x_1 - 2) = 0$$

Since $r \neq 0$ and $x_1 \neq 0$, then the last equality implies that $x_1 = 2$ and the right answer is $\boxed{\textbf{(B) } 2}$

Problem 2

Let g_n be a geometric sequence. Given that $g_5 = \frac{256}{3}$ and $g_8 = 36$. Find the 7-th term of the sequence.

(A) 12   (B) 18   (C) 24   (D) 48   (E) 64

Solution

Let g_1 be the first term and the r be the common ratio of the sequence. Then from the formula for the n-th term of a geometric sequence we have

$$g_5 = g_1 \cdot r^4$$
$$\frac{256}{3} = g_1 \cdot r^4$$

and

$$g_8 = g_1 \cdot r^7$$
$$36 = g_1 \cdot r^7$$

Dividing these equations we have

$$\frac{36}{\frac{256}{3}} = \frac{g_1 \cdot r^7}{g_1 \cdot r^4}$$
$$\frac{27}{64} = r^3$$
$$\frac{3}{4} = r$$

Therefore, we have

$$g_8 = g_7 \cdot r$$
$$36 = g_7 \cdot \frac{3}{4}$$
$$48 = g_7$$

and the right answer is $\boxed{\textbf{(D) } 48}$

Problem 3

The sum S of an infinite geometric series is equal to 400. Given that the second term of this series is equal to 75, find the product of all the possible common ratios of this sequence.

(A) 1 **(B)** $\frac{1}{4}$ **(C)** $\frac{3}{4}$ **(D)** $\frac{1}{16}$ **(E)** $\frac{3}{16}$

Solution

Let g_1 be the first term and the r be the common ratio of the sequence.

From the formula for the n-th term of a geometric sequence we have

$$g_2 = g_1 \cdot r$$
$$75 = g_1 \cdot r$$
$$\frac{75}{r} = g_1$$

Now from the formula for the sum of an infinite geometric series we have

$$S = \frac{g_1}{1 - r}$$
$$400 = \frac{g_1}{1 - r}$$
$$400(1 - r) = g_1$$

Substituting $g_1 = \frac{75}{r}$ into the last equation we have

$$400(1 - r) = g_1$$
$$400(1 - r) = \frac{75}{r}$$
$$400r(1 - r) = 75$$
$$400r - 400r^2 = 75$$
$$0 = 400r^2 - 400r + 75$$

Since the last equation is quadratic, then by the Vieta's Formulas[1] we have that the product of its zeros is

$$\frac{c}{a} = \frac{75}{400} = \frac{3}{16}$$

and the right answer is $\boxed{\textbf{(E)} \ \dfrac{3}{16}}$

---

[1]This topic is discussed in detail in Chapter 15 "Vieta's Formulas"

CHAPTER 29

TIME, SPEED, DISTANCE

A typical problem of the AMC 10 competition is a problem involving **time**, **speed** and **distance**. Knowing the relationship between these concepts is usually enough to solve most of the problems in this topic.

Let us assume that an object moves with a constant speed v and after the time t it covers some distance d. Then the following equalities are true and can be used interchangeably

$$d = v \cdot t$$
$$v = \frac{d}{t}$$
$$t = \frac{d}{v}$$

Let us now consider several examples.

Problem 1

Rania goes on a train to visit her grandmother. For some reason the train goes at $1/4$ of its usual speed and it takes an extra 24 minutes. What is the original time of the trip t in minutes when the train was going at a usual speed?

(A) 6   **(B)** 8   **(C)** 10   **(D)** 12   **(E)** 16

Solution

Let v be the slower speed of the train. Then the usual speed is $4v$. Notice that since the distance d is the same, then we can write the following equations

$$d = 4v \cdot t$$
$$d = v \cdot (t + 24)$$

This implies that

$$4v \cdot t = v \cdot (t + 24)$$
$$4\not{v} \cdot t = \not{v} \cdot (t + 24)$$
$$4t = t + 24$$

We can solve this equation by isolating t[1]

$$4t = t + 24$$
$$3t = 24$$
$$t = 8$$

and the right answer is $\boxed{\textbf{(B) } 8}$

Problem 2

Jorge walks at 4 mph and covers a certain distance d. If he walks at 6 mph, he covers the distance d and 2 miles more during the same time. What is the value of d?

(A) 2.4   **(B)** 2.8   **(C)** 3.2   **(D)** 3.6   **(E)** 4.0

Solution

Let t be the time. Notice that since the time is the same in both cases, then we can write two equations that include the time t and the distance d.

---

[1]This technique is discussed in Chapter 20 "Linear Equations. Part 1"

We have

$$t = \frac{d}{4}$$

$$t = \frac{d+2}{6}$$

This implies that

$$\frac{d}{4} = \frac{d+2}{6}$$

After cross-multiplying the terms we have

$$6d = 4(d+2)$$

This equation can be solved by isolating d^2

$$6d = 4(d+2)$$
$$6d = 4d + 8$$
$$2d = 8$$
$$d = 4$$

and the right answer is (E) 4.0

Problem 3

Elsa drove from her house to the supermarket at a speed of 40 kilometers per hour and back at a speed of 60 kilometers per hour. What was her average velocity?

(A) 45   (B) 48   (C) 50   (D) 52   (E) 55

Solution

Let d be the distance from the house to the supermarket. The total distance traveled, therefore, is $2d$.

Notice that the time traveled from the house to the supermarket can be found as

$$\frac{d}{40}$$

and the time traveled from the supermarket to the house can be found as

$$\frac{d}{60}$$

[2] This technique is discussed in Chapter 20 "Linear Equations. Part 1"

The total time traveled is therefore

$$\frac{d}{40} + \frac{d}{60} = \frac{d}{24}$$

The average velocity can be found as the total distance $2d$ divided by the total time $\frac{d}{24}$

$$\frac{2d}{\frac{d}{24}} = \frac{2d}{1} \cdot \frac{24}{d} = \frac{2\cancel{d}}{1} \cdot \frac{24}{\cancel{d}} = 48$$

and the right answer is $\boxed{\textbf{(B) } 48}$

CHAPTER 30

LINES ON A COORDINATE PLANE

The *Standard Form* of the equation of a line is

$$ax + by = c$$

where a, b, c are given real numbers, such that a and b are not simultaneously zero.

This equation produces horizontal lines when $a = 0$ and vertical lines when $b = 0$.

When $b \neq 0$ it can be rewritten in the *Slope-Intercept Form*

$$y = mx + b$$

The equation of the line with slope m that passes through the point (x_1, y_1) is given by the *Point-Slope Formula*

$$y - y_1 = m(x - x_1)$$

If m_1 and m_2 are the slopes of two given lines, then the lines are parallel when $m_1 = m_2$ and perpendicular when $m_1 \cdot m_2 = -1$.

Let us now consider several examples.

Problem 1

The lines $y = 2021x$ and $y = 21x + 20$ intersect at the point M. Which of the following intervals contains the square of the distance from the point M to the origin?

(A) $[0, 20]$ **(B)** $[40, 200]$ **(C)** $[220, 310]$ **(D)** $[320, 390]$ **(E)** $[400, 500]$

Solution

Let us start by finding the x-coordinate of the point of intersection of the lines[1]

$$2021x = 21x + 20$$
$$2000x = 20$$
$$x = \frac{20}{2000}$$
$$x = \frac{1}{100}$$

the y-coordinate of the point of intersection is

$$y = 2021 \cdot \frac{1}{100} = \frac{2021}{100}$$

Therefore, the square of the distance to the origin is equal to

$$x^2 + y^2 = \left(\frac{1}{100}\right)^2 + \left(\frac{2021}{100}\right)^2 = \frac{1 + 2021^2}{10000}$$

Notice that we can estimate[2] this value as follows

$$\frac{1 + 2021^2}{10000} > \frac{0 + 2000^2}{10000} = 400$$

and, therefore, the right answer is $\boxed{\textbf{(E)} \ [400, 500]}$

Problem 2

Nirvana is trying to draw a line perpendicular to the line $20px - 2000y = 2021$ and passing through the point $(20, 21)$. Her math teacher told her that the y-intercept of the new line is equal to 2021. For which values of p is this possible?

(A) -20 **(B)** -1 **(C)** 1 **(D)** 21 **(E)** $\frac{20}{21}$

---
[1] This equation is equivalent to a linear equation. You can find more problems that result in similar equations in Chapter 20 "Linear Equations. Part 1"
[2] This technique is discussed in Chapter 25 "Estimations"

Solution

Let us start by finding the slope of the given line

$$20px - 2000y = 2021$$
$$-2000y = -20px + 2021$$
$$y = \frac{20p}{2000}x - \frac{2021}{2000}$$
$$y = \frac{p}{100}x - \frac{2021}{2000}$$

From here the slope of the perpendicular line is the negative reciprocal of $\frac{p}{100}$, which equals to $-\frac{100}{p}$.

Now we will find the equation of the line passing through the point $(20, 21)$ with the slope $-\frac{100}{p}$. From the Point-Slope Formula we have

$$y - y_1 = m(x - x_1)$$
$$y - 21 = -\frac{100}{p}(x - 20)$$
$$y = -\frac{100}{p}(x - 20) + 21$$

From here by substituting $x = 0$ we can find the y-intercept to be

$$y = -\frac{100}{p}(0 - 20) + 21 = \frac{2000}{p} + 21$$

Since the y-intercept of the new line is equal to 2021, then we have the following equation

$$\frac{2000}{p} + 21 = 2021$$
$$\frac{2000}{p} = 2000$$
$$2000 = 2000p$$
$$1 = p$$

and the right answer is $\boxed{\textbf{(C) } 1}$

Problem 3

Find the product of all values of the real numbers a, such that the line

$$(a - 1)x + (a + 4)y = 10$$

cuts a right triangle of area 1 from the coordinate axes.

(A) 46 **(B)** -54 **(C)** 50 **(D)** 100 **(E)** -168

Solution

Let us start by finding the intercepts of the given line. For x-intercept we put $y = 0$ and we have

$$(a - 1)x + (a + 4)y = 10$$
$$(a - 1)x + (a + 4)(0) = 10$$
$$(a - 1)x = 10$$
$$x = \frac{10}{a - 1}$$

For y-intercept we put $x = 0$ and we have

$$(a - 1)x + (a + 4)y = 10$$
$$(a - 1)(x) + (a + 4)(y) = 10$$
$$(a + 4)y = 10$$
$$y = \frac{10}{a + 4}$$

Therefore, we have the following condition for the area

$$\frac{x \cdot y}{2} = 1$$
$$x \cdot y = 2$$
$$\frac{10}{a - 1} \cdot \frac{10}{a + 4} = 2$$
$$\frac{100}{(a - 1)(a + 4)} = 2$$
$$100 = 2(a - 1)(a + 4)$$
$$50 = (a - 1)(a + 4)$$
$$50 = a^2 + 3a + 4$$
$$0 = a^2 + 3a - 54$$

From here by Vieta's Formulas[3] we have that the product of the numbers a that satisfy this equation is equal to -54 and the right answer is $\boxed{\textbf{(B)} \ -54}$

---

[3]This topic is discussed in detail in Chapter 15 "Vieta's Formulas"

CHAPTER 31

UNITS DIGIT OF A NUMBER

The **units digit** of an integer number usually refers to its last digit. When dealing with the units digits it is very useful to keep in mind the following rule: the last digit of the result of such operations as addition, subtraction, multiplication, squaring, cubing (or other natural exponentiation) can be found by performing the operations only on the last digits of the numbers. This is equivalent to directly working in modulus 10.

Let us consider several examples.

Problem 1

Find the units digit of the number

$$123^{321}$$

(A) 1 **(B)** 3 **(C)** 5 **(D)** 7 **(E)** 9

Solution

Let us write the first powers of 3

$$3^1 = 3$$
$$3^2 = 9$$
$$3^3 = 27$$
$$3^4 = 81$$
$$3^5 = 243$$
$$3^6 = 729$$
$$3^7 = 2187$$
$$3^8 = 6561$$

Notice that the last digit follows the pattern "$3 - 9 - 7 - 1$" with a period of length 4. Since 321 gives remainder 1 when divided by 4, then 123^{321} has the same last digit as 3^1, which is 3 and the right answer is $\boxed{\textbf{(B) } 3}$

Problem 2

Find the units digit of the number

$$p = 2018^{2019} + 2019^{2020}$$

(A) 1   **(B)** 2   **(C)** 3   **(D)** 4   **(E)** 5

Solution

Let us find the last digit of the numbers 2018^{2019} and 2019^{2020} separately. For this we will look into the last digits of the powers of the numbers 8 and 9.

Let us write the first powers of 8

$$8^1 = 8$$
$$8^2 = 64$$
$$8^3 = 512$$
$$8^4 = 4096$$
$$8^5 = 32768$$

Notice that the last digit follows the pattern "$8 - 4 - 2 - 6$" with a period of length 4. Since 2019 gives remainder 3 when divided by 4, then 2018^{2019} has the same last digit as 8^3, which is 2.

Let us write the first powers of 9

$$9^1 = 9$$
$$9^2 = 81$$
$$9^3 = 729$$
$$9^4 = 6561$$

Notice that the last digit follows the pattern "9 − 1" with a period of length 2. Since 2020 gives remainder 0 when divided by 2, then 2019^{2020} has the same last digit as 9^2, which is 1.

Therefore, the last digit of the number p is $2 + 1 = 3$ and the right answer is $\boxed{\textbf{(C) } 3}$

Problem 3

Let $n = 4^{77} + 77^4$. Find the units digit of the number

$$(n + 4)^{n+77}$$

(A) 0   **(B)** 1   **(C)** 5   **(D)** 6   **(E)** 9

Solution

Let us write the first powers of 4

$$4^1 = 4$$
$$4^2 = 16$$
$$4^3 = 64$$
$$4^4 = 256$$
$$4^5 = 1024$$
$$4^6 = 4096$$

Notice that the last digit follows the pattern "4 − 6" with a period of length 2. Since 77 gives remainder 1 when divided by 2, then 4^{77} has the same last digit as 4^1, which is 4.

Since $7^4 = 2401$, then 77^4 ends on 1. From here we have that the last digit of the number n is 5 and the last digits of the number $n + 4$ is 9.

Let us now write the first powers of 9

$$9^1 = 9$$
$$9^2 = 81$$
$$9^3 = 729$$
$$9^4 = 6561$$

Notice that the last digit follows the pattern "$9 - 1$" with a period of length 2. Since n ends on 5, then n is odd and $n + 77$ is even. Therefore, the last digit of $(n+4)^{n+77}$ is 9 and the right answer is $\boxed{\textbf{(E) } 1}$

CHAPTER 32

TENS DIGIT OF A NUMBER

The **tens digit** of an integer number usually refers to its second last (or penultimate) digit. When dealing with the tens digits it is very useful to try and notice any obvious patterns or directly work in modulus 100.

Let us consider several examples.

Problem 1

Find the tens digit of the number
$$5^{2020}$$

(A) 0 **(B)** 1 **(C)** 2 **(D)** 3 **(E)** 5

Solution

Let us write the first powers of 5 and look for any obvious patterns.

We have

$$5^1 = 5$$
$$5^2 = 25$$
$$5^3 = 125$$
$$5^4 = 625$$

Notice that if the power of 5 is greater than 1, then the last two digits are always 25 and the right answer is $\boxed{(\textbf{C}) \ 2}$

Problem 2

Determine the tens digit of the number

$$11^{32}$$

(**A**) 2   (**B**) 3   (**C**) 4   (**D**) 5   (**E**) 6

Solution

Let us write the first powers of 11

$$11^1 = 11$$
$$11^2 = 121$$
$$11^3 = 1331$$
$$11^4 = 14641$$

Notice that for these numbers the last digit is always 1 and the tens digit is equal to the power. Let us show that the tens digit of 11^n equals the remainder of n modulo 10. Indeed the long multiplication of a number that ends with 1 by 11 consists of writing the original number, then writing it again shifted one position to the left, and then adding the results

	...	a	b	1
...	a	b	1	
...	...	...	$(b+1)$	1

From this process we notice that after each multiplication the tens digit is increased by 1. Therefore, the tens digit of 11^n equals the remainder of the power n modulo 10, which is 2 and the right answer is $\boxed{(\textbf{A}) \ 2}$

Problem 3

Find the tens digit of the number
$$7^{2021}$$

(A) 0 **(B)** 1 **(C)** 3 **(D)** 4 **(E)** 9

Solution

Let us write the first powers of 7

$$7^1 = 7$$
$$7^2 = 49$$
$$7^3 = 343$$
$$7^4 = 2401$$

Notice that 7^4 is congruent to 1 modulo 100. This means that the last two digits of this sequence follow the pattern "$07 - 49 - 43 - 01$" with a period of length 4. Since 2021 gives remainder 1 when divided by 4, then 7^{2021} has the same last two digits as 7^1, which are 07 and the right answer is $\boxed{\textbf{(A) } 0}$

CHAPTER 33

DECIMAL REPRESENTATION

The **decimal representation** of an integer number usually refers to the representation of the number using its digits in the base 10.

For example, the number 15472 can be written as

$$15472 = 10^4 \cdot 1 + 10^3 \cdot 5 + 10^2 \cdot 4 + 10^1 \cdot 7 + 10^0 \cdot 2$$

In many problems even though the number is unknown, it can still be written in its decimal representation. For example, the decimal representations for the two-digit number $\overline{ab}$ and the three-digit number $\overline{abc}$ are

$$\overline{ab} = 10a + b$$
$$\overline{abc} = 100a + 10b + c$$

Let us now consider several examples of this useful representation.

Problem 1

Jerry wrote some two-digit number x on the board. After this he wrote on the board the sum and the product of the digits of the number x. He noticed that if he adds the two new numbers he will obtain x. What is the units digit of x?

(A) 6   **(B)** 7   **(C)** 8   **(D)** 9   **(E)** 1

Solution

Notice that the two-digit number $x = \overline{ab}$ with the units digit a and the tens digit b can be written as
$$\overline{ab} = 10a + b$$
Therefore, we obtain the following equation
$$10a + b = ab + a + b$$
$$10a = ab + a$$
$$9a - ab = 0$$
$$a(9 - b) = 0$$

Since $a \neq 0$, then b should equal 9 and the right answer is $\boxed{\textbf{(D)}\ 9}$

Problem 2

How many three-digit numbers are exactly 90 times greater than the sum of their digits?

(A) 0   **(B)** 1   **(C)** 2   **(D)** 3   **(E)** 4

Solution

Notice that the three-digit number $\overline{abc}$ can be written as
$$\overline{abc} = 100a + 10b + c$$
Therefore, we obtain the following equation
$$100a + 10b + c = 90(a + b + c)$$
$$100a + 10b + c = 90a + 90b + 90c$$
$$10a - 80b = 89c$$
$$10(a - 8b) = 89c$$

Notice that the left-hand side is divisible by 10 and, therefore, so should be the right-hand side[1]. This means that $c = 0$ and, therefore, $a = 8b$.

[1] This technique is discussed in detail in Chapter 36 "Divisibility"

If $b = 0$, then $a = 0$ and we have do not have a three-digit number.

If $b = 1$, then $a = 8$ and the three-digit number is 810.

If $b \geq 2$, then $a = 8b \geq 16$ cannot be a digit of a number.

Therefore, there is only one such number and the right answer is $\boxed{(\textbf{B})\ 1}$

Problem 3

A two-digit number is called *awesome* if when divided by the sum of its digits, its quotient is 3 and its remainder is 2. How many *awesome* numbers are there?

(**A**) 1   (**B**) 3   (**C**) 5   (**D**) 7   (**E**) 9

Solution

Notice that the two-digit number $\overline{ab}$ can be written as

$$\overline{ab} = 10a + b$$

Therefore, we obtain the following equation

$$10a + b = 3(a + b) + 2$$

which is equivalent to

$$10a + b = 3a + 3b + 2$$
$$7a = 2b + 2$$

Notice that since b is a digit, then $2b + 2 \leq 2 \cdot 9 + 2 = 20$.

This implies that $7a \leq 20$ and the only values that we need to check are $a = 0$, $a = 1$ and $a = 2$.

If $a = 0$, then the equation $0 = 2b + 2$ implies that $b = -1$, which cannot be a digit.

If $a = 1$, then the equation $7 = 2b + 2$ implies that $b = \frac{5}{2}$, which cannot be a digit.

If $a = 2$, then the equation $14 = 2b + 2$ implies that $b = 6$ and we have an *awesome* number 26.

Therefore, there is only one *awesome* number and the right answer is $\boxed{(\textbf{A})\ 1}$

CHAPTER 34

REPEATING DECIMALS

A real number is a **repeating decimal** if after certain time it becomes strictly periodic. For the notation purposes, the period usually has a horizontal line above it. For example

$$0.333333... = 0.\overline{3}$$

$$-2.545454... = -2.\overline{54}$$

$$9.2761761... = 9.2\overline{761}$$

A finite decimal number is called a *terminating decimal*. A fraction can be converted to a repeating or terminating decimal form by simply applying the division algorithm. Terminating and repeating decimal can be converted to a fractional form by the following rules.

1. If the terminating decimal is of the form

$$a_1 a_2 ... a_k . b_1 b_2 ... b_m$$

then it equals to the fraction

$$a_1 a_2 ... a_k + \frac{1}{10^m} (b_1 b_2 ... b_m)$$

2. If the repeating decimal is of the form

$$a_1 a_2 ... a_k . \overline{c_1 c_2 ... c_n}$$

then it equals to the fraction

$$a_1 a_2 ... a_k + \frac{c_1 c_2 ... c_n}{99...9}$$

where the digit 9 repeats n times.

3. If the repeating decimal is of the form

$$a_1 a_2 ... a_k . b_1 b_2 ... b_m \overline{c_1 c_2 ... c_n}$$

then it equals to the fraction

$$a_1 a_2 ... a_k + \frac{1}{10^m} \left(b_1 b_2 ... b_m + \frac{c_1 c_2 ... c_n}{99...9} \right)$$

where the digit 9 repeats n times.

For example

$$7.125 = 7 + \frac{1}{10^3}(125) = 7 + \frac{1}{8} = \frac{57}{8}$$

$$3.\overline{18} = 3 + \frac{18}{99} = 3 + \frac{2}{11} = \frac{35}{11}$$

$$0.2\overline{345} = 0 + \frac{1}{10^1}\left(2 + \frac{345}{999} \right) = \frac{781}{3330}$$

Let us now consider several problems that involve repeating decimals.

Problem 1

Given the repeating decimals numbers $A = 1.222222...$ and $B = 2.343434....$ Which of the following equals to $B - A$?

(A) $\frac{1}{33}$ (B) $\frac{1}{99}$ (C) $\frac{4}{33}$ (D) $\frac{37}{33}$ (E) $\frac{110}{99}$

Solution

Let us start by converting the numbers A and B to fractional form

$$A = 1.222222... = 1.\overline{2} = 1 + \frac{2}{9} = \frac{11}{9}$$

$$B = 2.343434... = 2.\overline{34} = 2 + \frac{34}{99} = \frac{232}{99}$$

Now we can find the value of $B - A$ as

$$B - A = \frac{232}{99} - \frac{11}{9} = \frac{37}{33}$$

and the right answer is $\boxed{\textbf{(D)} \ \dfrac{37}{33}}$

Problem 2

The repeating decimal number $x = 0.\overline{ab}$ is m times greater than the terminating decimal number $y = 0.ab$. Which of the intervals contains the value of m?

(A) $(0, 0.1)$ **(B)** $(0, 0.9)$ **(C)** $(0.9, 0.99)$ **(D)** $(1, 1.01)$ **(E)** $(1.01, 1.1)$

Solution

We will start by writing the equation equivalent to the statement that x is m times greater than y

$$x = m \cdot y$$
$$0.\overline{ab} = m \cdot 0.ab$$
$$\frac{ab}{99} = m \cdot \frac{1}{10^2}(ab)$$
$$\frac{\cancel{ab}}{99} = m \cdot \frac{1}{10^2}(\cancel{ab})$$
$$\frac{1}{99} = m \cdot \frac{1}{100}$$
$$\frac{100}{99} = m$$

However

$$\frac{100}{99} = 1 + \frac{1}{99} = 1.\overline{01} = 1.010101...$$

and, therefore, the right answer is $\boxed{\textbf{(E)} \ (1.01, 1.1)}$

Problem 3

The repeating decimal number $0.\overline{abc}$ with period 3 is written as a fraction in the simplest form $\frac{m}{n}$. Find the number of all possible values of n.

(A) 4 **(B)** 5 **(C)** 7 **(D)** 8 **(E)** 10

Solution

We will start by writing the repeating decimal number $0.\overline{abc}$ as a fraction

$$0.\overline{abc} = \frac{abc}{999}$$

It is now clear that the possible denominators n should divide 999. The positive integer divisors of 999 are 1, 3, 9, 27, 37, 111, 333, 999.

If the denominator is 1, then $\overline{abc}$ has a factor of 999. This means that $\overline{abc} = 999$ and does not have period 3.

If the denominator is 3, then $\overline{abc}$ has a factor of 333. This means that $\overline{abc}$ is either 333, 666 or 999 and does not have period 3.

If the denominator is 9, then $\overline{abc}$ has a factor of 111. This means that $\overline{abc}$ is either 111, 222, ... , 999 and does not have period 3.

It is not hard to see that all other denominators can represent a repeating decimal with period 3. Indeed, we have

$$\frac{1}{27} = \frac{37}{999} = 0.\overline{037}$$

$$\frac{1}{37} = \frac{27}{999} = 0.\overline{027}$$

$$\frac{1}{111} = \frac{9}{999} = 0.\overline{009}$$

$$\frac{1}{333} = \frac{3}{999} = 0.\overline{003}$$

$$\frac{1}{999} = 0.\overline{001}$$

Therefore, there are 5 such values of n and the right answer is $\boxed{\textbf{(B) } 5}$

CHAPTER 35

NUMBER BASES

Let us consider a fixed positive integer number $b > 1$, which we will call the **base**. For a given a positive integer number n, we will consider an m-tuple of nonnegative integer numbers

$$(a_m, a_{m-1}, ..., a_1, a_0)_b$$

such that

$$n = a_m b^m + a_{m-1} b^{m-1} + ... + a_1 b^1 + a_0 b^0$$

where $a_i < b$, $a_m \neq 0$.

Such m-tuple is unique and is called the **representation of n in base b**.

We can use this definition to convert the numbers from any base to the usual base 10. For example, if the number n is given in base 7 as $n = (523)_7$, then we can convert it to decimal system as follows

$$n = (523)_7 = 5 \cdot 7^2 + 2 \cdot 7^1 + 3 \cdot 7^0 = 262$$

We can use the algorithm of division to convert the numbers from the base 10 to any other base. For example, if the number n is given as 262 in the decimal system, then we can divide it by 7 several times and arrange the remainders of the division as a representation in base 7. Start by dividing 262 by 7

$$262 = 7 \cdot 37 + 3$$

This means that 3 is going to be the last digit in base 7. Now divide 37 by 7

$$37 = 7 \cdot 5 + 2$$

This means that 2 is going to be the next digit in base 7. Since 5 cannot be divided by 7, then the remaining digit is 5 and the result is

$$(523)_7$$

Let us now consider several problems that involve numbers in different bases.

Problem 1

Let $(\overline{xy})_b$ represent a two-digit number $\overline{xy}$ written in base b. Find the value of b, for which $(51)_b$ is three times greater than $(15)_b$.

(A) 5   (B) 6   (C) 7   (D) 8   (E) 9

Solution

We will start by writing the value of both numbers

$$(51)_b = 5 \cdot b + 1 = 5b + 1$$
$$(15)_b = 1 \cdot b + 5 = b + 5$$

From here we have the following equation[1]

$$5b + 1 = 3(b + 5)$$
$$5b + 1 = 3b + 15$$
$$2b = 14$$
$$b = 7$$

and the right answer is (C) 7

[1]This equation is equivalent to a linear equation. You can find more problems that result in similar equations in Chapter 20 "Linear Equations. Part 1"

Problem 2

How many three-digit numbers b are there, such that the number 2021 written in the base b ends with the digit 1?

(A) 4   **(B)** 6   **(C)** 8   **(D)** 12   **(E)** 16

Solution

Since 2021 written in the base b ends with the digit 1, then we have

$$2021 = a_m b^m + a_{m-1} b^{m-1} + ... + a_1 b^1 + 1$$

or equivalently

$$2020 = b \cdot \left(a_m b^{m-1} + a_{m-1} b^{m-2} + ... + a_1 \right)$$

This implies that b divides 2020. From the prime factorization of 2020

$$2020 = 2^2 \cdot 5 \cdot 101$$

The only three digit divisors of 2020 are 101, 202, 404 and 505. Therefore, there are only 4 such values of b and the right answer is $\boxed{\textbf{(A)} \, 4}$

Problem 3

Which of the following numbers is divisible by 6 in any base larger than 3?

(A) 1020   **(B)** 1032   **(C)** 1230   **(D)** 1320   **(E)** 3102

Solution

Let us show that the number 1320 is divisible by 6 in any base $b > 3$. We have

$$(1320)_b = 1 \cdot b^3 + 3 \cdot b^2 + 2 \cdot b^1 + 0 \cdot b^0$$
$$(1320)_b = b^3 + 3b^2 + 2b$$
$$(1320)_b = b(b+1)(b+2)$$

Notice that among any three consecutive numbers b, $b+1$ and $b+2$, there is always a number divisible by 2 and a number divisible by 3. This implies that the number $(1320)_b$ is divisible by 6 for all b and the right answer is $\boxed{\textbf{(D)} \, 1320}$

CHAPTER 36

DIVISIBILITY

Let us assume that we are given two integer numbers n and m, $m \neq 0$. We will say that the number n is divisible by m if there exists an integer number k, such that

$$n = m \cdot k$$

The number m in this case is called the *divisor* of the number n and the following notation is used

$$m \mid n$$

This notation is usually read as "*m divides n*".

For example, 15 is divisible by 3, because there exists an integer number 5 that satisfy the equality

$$15 = 3 \cdot 5$$

The **divisibility** of two integer numbers is an important property used in many number theory problems. Let us now consider several examples that involve the concept of divisibility.

Problem 1

Avner wrote a 6-digit number divisible by 9 on the board. Find the remainder when this number is divided by 5 if the first five digits of the number are 1, 2, 3, 4 and 5 in some order.

(A) 0   **(B)** 1   **(C)** 2   **(D)** 3   **(E)** 4

Solution

Let the last digit of the number be x. Notice that any integer number always gives the same remainder modulo 9 as the sum of its digits. The sum of the digits of the number is

$$1 + 2 + 3 + 4 + 5 + x = 15 + x$$

Since x ranges from 0 to 9, then $15 + x$ ranges from 15 to 24 and the only number divisible by 9 in that interval is 18. This implies that $x = 3$. Since the number ends on 3, then its remainder modulo 5 is equal to 3 and the right answer is $\boxed{\textbf{(D)} \ 3}$

Problem 2

Which of the following integers cannot be written as a sum of seven consecutive integers?

(A) 1456   **(B)** 2177   **(C)** 2884   **(D)** 3590   **(E)** 4263

Solution

Let the seven consecutive numbers be

$$x, x + 1, x + 2, x + 3, x + 4, x + 5, x + 6$$

and the sum of the numbers is

$$x + (x + 1) + (x + 2) + (x + 3) + (x + 4) + (x + 5) + (x + 6) = 7x + 21$$

Notice that this number is divisible by 7. The only number in the list that is not divisible by 7 is 3590 and, therefore, the right answer is $\boxed{\textbf{(D)} \ 3590}$

Problem 3

Given a number of the form $N = \overline{abcabc}$, where a, b and c are some non-zero digits. Which of the following numbers is a divisor of all such numbers?

(A) 13   **(B)** 17   **(C)** 22   **(D)** 26   **(E)** 101

Solution

Notice that the number N can be written as

$$N = \overline{abcabc} = \overline{abc} \cdot 1000 + \overline{abc} = 1001 \cdot \overline{abc}$$

The prime factorization of 1001 is

$$N = 7 \cdot 11 \cdot 13$$

The only number from the list that is a divisor of all such numbers is 13 and, therefore, the right answer is $\boxed{\textbf{(A) } 13}$

CHAPTER 37

REMAINDERS

In many AMC 10 problems we have to work with the integer numbers and the **remainders** of their division by other integers.

Let us assume that we are given two integer numbers n and m, $m \neq 0$. If the result of the division is k and the remainder is r, $0 \leq r \leq m - 1$, then the number n can be written in the following form

$$n = k \cdot m + r$$

For example, if we divide the number 43 by 5 we will get 8 as a result and 3 as a remainder of the division. The number 43 in this case can be written as

$$43 = 5 \cdot 8 + 3$$

Let us now consider several examples that involve remainders.

Problem 1

How many numbers from 1 to 2016 inclusively give remainder 1 when divided by 2 and give remainder 2 when divided by 3?

(A) 334   **(B)** 335   **(C)** 336   **(D)** 337   **(E)** 338

Solution

Notice that since 2 and 3 are relatively prime, then we can work modulo 6. The numbers from 1 to 2016 will be of one of the following forms

$$6k, 6k + 1, 6k + 2, 6k + 3, 6k + 4, 6k + 5$$

The numbers of the form $6k$, $6k + 2$ and $6k + 4$ are even and, therefore, cannot give remainder 1 when divided by 2. The numbers of the form $6k + 3$ are divisible by 3 and, therefore, cannot give remainder 2 when divided by 3. The numbers of the form $6k + 1$ give remainder 1 when divided by 3. Therefore, this implies that only the numbers of the form $6k + 5$ satisfy the conditions of the problem.

Since there are

$$\frac{2016}{6} = 336$$

such numbers, then the right answer is $\boxed{\textbf{(C)} \ 336}$

Problem 2

Which of the following intervals contains the number that gives remainder 3 when divided by 5, 8 and 10?

(A) $[22, 28]$ **(B)** $[28, 34]$ **(C)** $[34, 40]$ **(D)** $[40, 46]$ **(E)** $[46, 52]$

Solution

Notice that the least common multiple of the numbers 5, 8 and 10 equals 40. The number that will give remainder 3 when divided by 5, 8 and 10 will also give the same remainder when divided by 40. Therefore, the needed number is of the form

$$40k + 3$$

The only such number not greater than 52 is 43 and, therefore, the right answer is $\boxed{\textbf{(D)} \ [40, 46]}$

Problem 3

Find the number of three digit numbers that give remainder 1 when divided by 10 and remainder 2 when divided by 11?

(A) 6   **(B)** 7   **(C)** 8   **(D)** 9   **(E)** 10

Solution

Notice that since 10 and 11 are relatively prime, then we can divide our numbers into blocks of 110 numbers each. Among the numbers from 1 to 110 inclusively only the following numbers give remainder 1 when divided by 10

$$11, 21, 31, 41, 51, 61, 71, 81, 91, 101$$

Among these numbers only the number 101 gives remainder 2 when divided by 11. Therefore, in each block of 110 consecutive integers there is only one number that satisfies the conditions of the problem. It is not hard to see that there are only 9 such numbers less than 1000

$$101, 211, 321, 431, 541, 651, 761, 871, 981$$

and the right answer is $\boxed{\textbf{(D)} \ 9}$

CHAPTER 38

MODULUS

In many AMC 10 problems we have to consider the remainders of the given integer numbers when divided by a certain number called **modulus**.

Let us now consider several examples that use this idea.

Problem 1

Which of the following represents the number of integer solutions (n, k) of the equation

$$n^2 = 3k + 2$$

(A) 0   **(B)** 1   **(C)** 2   **(D)** 4   **(E)** 8

Solution

We will work modulus 3. Notice that the right-hand side of the equation gives remainder 2 when divided by 3. Let us, therefore, consider the possible remainders of the expression n^2.

If $n \equiv 0 \pmod 3$, then
$$n^2 \equiv (0)^2 \equiv 0 \pmod 3$$
If $n \equiv 1 \pmod 3$, then
$$n^2 \equiv (1)^2 \equiv 1 \pmod 3$$
If $n \equiv 2 \pmod 3$, then
$$n^2 \equiv (2)^2 \equiv 1 \pmod 3$$
We can see that n^2 always gives remainder 0 or 1 when divided by 3. Therefore, there are no such integer numbers and the right answer is $\boxed{\textbf{(A)}\ 0}$

Problem 2

Determine the number of integer numbers a, that satisfy the equation
$$(a^2)! + 1 = 2b^3$$
for some integer value of b.

(A) 0   **(B)** 1   **(C)** 2   **(D)** 3   **(E)** 4

Solution

We will work modulus 2. Notice that the right-hand side of the equation is always an even number. Let us, consider the parity of the expression on the left-hand side.

First, let us consider the case when $|a| \geq 2$. The factorial of the number a^2 will contain a factor of 2 and thus will be an even number, which means that the left-hand side is odd, while the right-hand side is even. Therefore, there are no solutions in this case.

Now let us consider the case when $|a| < 2$. We only need to consider three cases: $a = -1$, $a = 0$ and $a = 1$. Since $0! = 1$ and $1! = 1$, then for each value $a = -1$, $a = 0$ and $a = 1$ we have integer solutions of the equation
$$((0)^2)! + 1 = 2(1)^3$$
$$((-1)^2)! + 1 = 2(1)^3$$
$$((1)^2)! + 1 = 2(1)^3$$
Therefore, there are three such values of the number a and the right answer is $\boxed{\textbf{(D)}\ 3}$

Problem 3

Determine the sum of all positive integer numbers n and m, such that
$$2^{n-2021} = m^2 + 1$$

(A) 0   **(B)** 2020   **(C)** 2021   **(D)** 2022   **(E)** 2023

Solution

We will start by noticing that when $n < 2021$, the left-hand side is less than the right-hand side. Indeed

$$2^{n-2021} = \frac{1}{2^{2021-n}} < \frac{1}{2^1} < 1 < m^2 + 1$$

and, therefore, there are no positive integer solutions in this case.

Let us now solve the equation in positive integers when $n = 2021$. We have

$$2^{n-2021} = m^2 + 1$$
$$2^{(2021)-2021} = m^2 + 1$$
$$2^0 = m^2 + 1$$
$$0 = m^2$$
$$0 = m$$

and, therefore, there are no positive integer solutions in this case.

The equation, however, has positive integer solutions when $n = 2022$. Indeed

$$2^{n-2021} = m^2 + 1$$
$$2^{(2022)-2021} = m^2 + 1$$
$$2^1 = m^2 + 1$$
$$1 = m^2$$
$$1 = m$$

Now we will consider the case $n > 2022$. We will work modulus 4. Notice that the left-hand side of the equation is divisible by 4. Let us, therefore, consider the remainder of the expression on the right-hand side by modulus 4.

If $m \equiv 0 \pmod 4$, then

$$m^2 + 1 \equiv (0)^2 + 1 \equiv 1 \pmod 4$$

If $m \equiv 1 \pmod 4$, then

$$m^2 + 1 \equiv (1)^2 + 1 \equiv 2 \pmod 4$$

If $m \equiv 2 \pmod 4$, then

$$m^2 + 1 \equiv (2)^2 + 1 \equiv 1 \pmod 4$$

If $m \equiv 3 \pmod 4$, then

$$m^2 + 1 \equiv (3)^2 + 1 \equiv 2 \pmod 4$$

We can see that n^2 always gives remainder 1 or 2 when divided by 4. Therefore, there are no such integer numbers in this case.

The sum of all positive integer numbers n and m that satisfy the equation, therefore, is

$$m + n = 2022 + 1 = 2023$$

and the right answer is (E) 2023

CHAPTER 39

GREATEST COMMON DIVISOR

The **Greatest Common Divisor** of two integer numbers m and n (sometimes simply written as *GCD* or *gcd*) is the largest positive integer number d, such that d divides m and d divides n. The notation for the greatest common divisor of the numbers m and n is

$$\gcd(m, n)$$

For example, the greatest common divisor of the numbers 60 and 36 is 12, i.e.

$$\gcd(60, 36) = 12$$

Indeed, both numbers are divisible by 12 and there is no number larger than 12 that is a divisor of both 60 and 36.

The simplest way to find the greatest common divisor is to consider the prime factorization of the given numbers. Then the greatest common divisor will consist of their common prime divisors raised to the smallest corresponding present powers.

In our example

$$60 = 2^2 \cdot 3^1 \cdot 5$$
$$36 = 2^2 \cdot 3^2$$

The common prime divisors of these numbers are 2 and 3. The smallest power of the prime divisor 2 is 2 and the smallest power of the prime divisor 3 is 1. Therefore, we have

$$\gcd(60, 36) = 2^2 \cdot 3^1 = 12$$

Let us now consider several examples of the problems that involve the concept of the greatest common divisor.

Problem 1

The greatest common divisor of the numbers 276, 204 and x is 12. Which of the following intervals contains a possible value of x?

(A) $[49, 59]$ **(B)** $[74, 83]$ **(C)** $[91, 99]$ **(D)** $[109, 119]$ **(E)** $[123, 131]$

Solution

Note that x should be a multiple of 12. The only interval that contains a multiple of 12 is $[91, 99]$ (since 96 is divisible by 12). Therefore, the right answer is $\boxed{\textbf{(C)} [91, 99]}$

Problem 2

How many positive integer numbers N less than 100 are there, such that the greatest common divisor of the numbers $3N^2 + N - 19$ and $4N^2 - N + 86$ is equal to 7?

(A) 0   **(B)** 4   **(C)** 8   **(D)** 12   **(E)** 16

Solution

Note that $3N^2 + N - 19$ and $4N^2 - N + 86$ are both divisible by 7. Let us consider the sum of these numbers

$$\left(3N^2 + N - 19\right) + \left(4N^2 - N + 86\right) = 7N^2 + 67$$

This implies that $7N^2 + 67$ should also be divisible by 7. However, since $7N$ is divisible by 7, so should be 67, which is false[1]. Therefore, the greatest common divisor of $3N^2 + N - 19$ and $4N^2 - N + 86$ cannot be equal to 7 and there are no such integer numbers N. The right answer is $\boxed{\textbf{(A)} 0}$

---

[1] These types of arguments are discussed in detail in Chapter 36 "Divisibility"

Problem 3

Given the positive integers a, b, c, such that $c < a < b$. Which of the following is the greatest common divisor of the numbers 6^{a+b}, 10^{b+c} and 30^{c+a}.

(A) 2^{a+b} (B) 2^{c+a} (C) 2^{b+c} (D) 3^{a+b} (E) 5^{c+a}

Solution

Let us start by noticing that since $c < a < b$, then the following inequality is true

$$c + a < b + c < a + b$$

Let us rewrite the numbers as

$$6^{a+b} = 2^{a+b} \cdot 3^{a+b}$$
$$10^{b+c} = 2^{b+c} \cdot 5^{b+c}$$
$$30^{c+a} = 2^{c+a} \cdot 3^{c+a} \cdot 5^{c+a}$$

Therefore, the greatest common divisor should contain only the prime number 2 with the largest common power $c + a$ and the right answer is $\boxed{\textbf{(B) } 2^{c+a}}$

CHAPTER 40

LEAST COMMON MULTIPLE

The **Least Common Multiple** of two integer numbers m and n (sometimes simply written as *LCM* or *lcm*) is the smallest positive integer number l, such that m divides l and n divides l. The notation for the greatest common divisor of the numbers m and n is

$$\text{lcm}\,(m, n)$$

For example, the least common multiple of the numbers 12 and 40 is 120

$$\text{lcm}\,(12, 40) = 120$$

Indeed, 120 is divisible by both numbers and there is no smaller number that is divisible by both 12 and 40.

The simplest way to find the least common multiple is to consider the prime factorization of the given numbers. Then the least common multiple will consist of all the prime divisors raised to the largest corresponding present powers.

In our example

$$12 = 2^2 \cdot 3^1$$
$$40 = 2^3 \cdot 5^1$$

The prime divisors of these numbers are 2, 3 and 5. The largest power of the prime divisor 2 is 3, the largest power of the prime divisor 3 is 1 and the largest power of the prime divisor 5 is 1. Therefore, we have

$$\text{lcm}\,(12, 40) = 2^3 \cdot 3^1 \cdot 5^1 = 120$$

Let us now consider several examples of the problems that involve the concept of the least common multiple.

Problem 1

Let M be the least common multiple of the numbers 10^2, 4^3 and 6^3. What is the sum of digits of M?

(A) 1   (B) 3   (C) 5   (D) 7   (E) 9

Solution

Let us start by writing the prime factorization of the numbers 10^2, 4^3 and 6^3

$$10^2 = 2^2 \cdot 5^2$$
$$4^3 = 2^6$$
$$6^3 = 2^3 \cdot 3^3$$

Therefore, the greatest common divisor should contain the prime numbers 2, 3 and 5 with their largest present powers

$$2^6 \cdot 3^3 \cdot 5^2 = 43200$$

The sum of the digits of this number is

$$4 + 3 + 2 + 0 + 0 = 9$$

and the right answer is $\boxed{\textbf{(E)}\ 9}$

Problem 2

Johann found the least common multiple of two numbers and wrote it on the board: 180. Which of the following is a possible value of the product of these numbers?

(A) 120   (B) 240   (C) 360   (D) 420   (E) 560

Solution

Since the greatest common divisor multiplied by the least common multiple always equals to the product of the numbers

$$\gcd(x, y) \cdot \operatorname{lcm}(x, y) = x \cdot y$$

then the product of the numbers should be divisible by 180. The only such number in the list is 360 and the right answer is $\boxed{\textbf{(C)}\ 360}$

Problem 3

Given the positive integers a, b, c, such that $c < a < b$. Which of the following is the least common multiple of the numbers 6^{a+b}, 10^{b+c} and 30^{c+a}.

(A) 2^{c+a} (B) 3^{a+b} (C) 5^{b+c} (D) $5^{c+a} \cdot 6^{a+b}$ (E) $6^{a+b} \cdot 5^{b+c}$

Solution

Let us start by noticing that since $c < a < b$, then the following inequality is true

$$c + a < b + c < a + b$$

Let us rewrite the numbers as

$$6^{a+b} = 2^{a+b} \cdot 3^{a+b}$$
$$10^{b+c} = 2^{b+c} \cdot 5^{b+c}$$
$$30^{c+a} = 2^{c+a} \cdot 3^{c+a} \cdot 5^{c+a}$$

Therefore, the least common multiple should contain the prime number 2 with the power $a + b$, the prime number 3 with the power $a + b$ and the prime number 5 with the power $b + c$

$$\operatorname{lcm}\left(6^{a+b}, 10^{b+c}, 30^{c+a}\right) = 2^{a+b} \cdot 3^{a+b} \cdot 5^{b+c} = 6^{a+b} \cdot 5^{b+c}$$

and the right answer is $\boxed{\textbf{(E)}\ 6^{a+b} \cdot 5^{b+c}}$

CHAPTER 41

PROBABILITY

In many AMC 10 problems we are asked to find the probability of some event. The **simple probability** of an event can be found by dividing the number of outcomes of the event by the total number of possible outcomes.

For example, if we toss a coin and we are asked about the probability P to get a tail, then we can just divide 1 (the outcome of getting a *tail*) by 2 (total number of possible outcomes: *tail* or *head*)

$$P = \frac{1}{2}$$

Let us now consider several examples.

Problem 1

Enrique tosses a coin 3 times. Find the probability that he will get exactly two consecutive tails.

(A) $\frac{1}{8}$ (B) $\frac{1}{4}$ (C) $\frac{1}{2}$ (D) $\frac{2}{3}$ (E) $\frac{3}{4}$

Solution

The total number of possible sequences of heads and tails can be found as[1]

$$2^3 = 8$$

Notice that only the sequences TTH and HTT have two consecutive tails. Therefore, the probability that there are exactly two consecutive tails is equal to

$$P = \frac{2}{8} = \frac{1}{4}$$

and the right answer is $\boxed{\textbf{(B)}\ \dfrac{1}{4}}$

Problem 2

A fair die is rolled twice. What is the probability that the sum of the obtained numbers equals 4?

(A) $\frac{1}{2}$ **(B)** $\frac{1}{3}$ **(C)** $\frac{1}{9}$ **(D)** $\frac{1}{12}$ **(E)** $\frac{1}{18}$

Solution

The total number of possible outcomes is equal to

$$6 \cdot 6 = 36$$

Notice that only the numbers $(1,3)$, $(3,1)$ and $(2,2)$ add up to 4. Therefore, the probability that the sum of the obtained numbers equals 4 is

$$P = \frac{3}{36} = \frac{1}{12}$$

and the right answer is $\boxed{\textbf{(D)}\ \dfrac{1}{12}}$

Problem 3

Micah draws two cards simultaneously from a deck of 52 cards. Let the probability that he will get two black cards be written as a fraction $\frac{m}{n}$ in its simplest form. Which of the following equals to the value $m + n$?

(A) 5 **(B)** 8 **(C)** 53 **(D)** 127 **(E)** 155

---

[1]This result can be explained using the principle discussed in detail in Chapter 42 "Multiplication Principle"

Solution

Notice that the total number of all possible outcomes when drawing two cards simultaneously is equal to the number of ways to choose[2] two objects out of 52 objects

$$\binom{52}{2} = \frac{52 \cdot 51}{2} = 26 \cdot 51$$

Since there are 26 black cards in the deck, then the number of outcomes, where both cards are black is equal to the number of ways to choose two objects out of 26 objects

$$\binom{26}{2} = \frac{26 \cdot 25}{2} = 13 \cdot 25$$

Therefore, the probability that Micah will get two black cards is equal to

$$P = \frac{13 \cdot 25}{26 \cdot 51} = \frac{25}{102}$$

From here $m = 25$, $n = 102$, $m + n = 25 + 102 = 127$ and the right answer is

$\boxed{\textbf{(D) } 127}$

---

[2]This idea is discussed in detail in Chapter 45 "Combinations"

CHAPTER 42

MULTIPLICATION PRINCIPLE

The **Multiplication Principle** of counting states that if two events are independent, then the number of possible outcomes of these events, can be found by multiplying the outcomes of each event. The Multiplication Principle is usually applied when we work with events that do not depend on one another.

Let us now consider several examples.

Problem 1

Paola drew a table 2×2 consisting of four squares. Paola only has black, white and blue pencils. How many possible colorings of the table are there?

(A) 12   **(B)** 24   **(C)** 45   **(D)** 60   **(E)** 81

Solution

Notice that the coloring of each individual square is an independent event and has only 3 possibilities. Therefore, by the Multiplication Principle the number of possible colorings is equal to

$$3 \cdot 3 \cdot 3 \cdot 3 = 3^4 = 81$$

and the right answer is $\boxed{\textbf{(E)} \ 81}$

Problem 2

We call a natural number *odd-looking* if all of its digits are odd and the last digit is greater than 7. How many four-digit *odd-looking* numbers are there?

(A) 250   **(B)** 240   **(C)** 200   **(D)** 125   **(E)** 500

Solution

Notice that the choice of each digit is an independent event. The first three digits have 5 possibilities to choose from: 1, 3, 5, 7 or 9. The last digit has only 1 possibility to choose from: 9. Therefore, by the Multiplication Principle the number of four-digit *odd-looking* numbers is

$$5 \cdot 5 \cdot 5 \cdot 1 = 125$$

and the right answer is $\boxed{\textbf{(D)} \ 125}$

Problem 3

Let M be the number of 5-digit numbers that are even and contain no zero digits and N be the number of 5-digit numbers that are odd and do not contain digit 1. Which of the following is closer to the value of $M - N$?

(A) 2800   **(B)** 2900   **(C)** 3000   **(D)** 3100   **(E)** 3200

Solution

Notice that the choice of each digit is an independent event. Let us find the value of M. The first four digits have 9 possibilities to choose from: 1, 2, 3, 4, 5, 6, 7, 8 or 9. For the last digit we have only 4 possibilities to choose from: 2, 4, 6, 8. Therefore, by the Multiplication Principle the number M is

$$M = 9 \cdot 9 \cdot 9 \cdot 9 \cdot 4 = 4 \cdot 9^4$$

Let us now find the value of N. The first digit has 8 possibilities to choose from: 2, 3, 4, 5, 6, 7, 8 or 9. The next three digits have 9 possibilities to choose from: 0, 2, 3, 4, 5, 6, 7, 8 or 9. For the last digit we have only 4 possibilities to choose from: 3, 5, 7, 9. Therefore, by the Multiplication Principle the number N is

$$M = 8 \cdot 9 \cdot 9 \cdot 9 \cdot 4 = 4 \cdot 8 \cdot 9^3$$

Therefore, the value of $M - N$ is

$$M - N = 4 \cdot 9^4 - 4 \cdot 8 \cdot 9^3 = 2916$$

and the right answer is $\boxed{\textbf{(B)}\ 2900}$

CHAPTER 43

ADDITION PRINCIPLE

The **Addition Principle** of counting states that if two events are mutually exclusive, then the number of possible outcomes of these events, can be found by adding the outcomes of each event. The Addition Principle is usually applied when we are required to do some casework.

Let us now consider several examples.

Problem 1

How many three-digit numbers consist of distinct digits and have exactly one 9?

(**A**) 100   (**B**) 128   (**C**) 200   (**D**) 256   (**E**) 512

Solution

Notice that there are three possibilities for the position of the digit 9: it can be hundreds digit $\overline{9ab}$, tens digit $\overline{a9b}$ or units digit $\overline{ab9}$. We will consider these three cases apart.

Case 1: the number is of the form $\overline{9ab}$. The digit a can take any value from 0 to 8 and, therefore, has 9 possibilities. The digit b cannot equal a and, therefore, has 8 possibilities. By the Multiplication Principle the number of such numbers is equal to

$$9 \cdot 8 = 72$$

Case 2: the number is of the form $\overline{a9b}$. The digit a can take any value from 1 to 8 and, therefore, has 8 possibilities. The digit b cannot equal a and, therefore, has 8 possibilities. By the Multiplication Principle the number of such numbers is equal to

$$8 \cdot 8 = 64$$

Case 3: the number is of the form $\overline{ab9}$. The digit a can take any value from 1 to 8 and, therefore, has 8 possibilities. The digit b cannot equal a and, therefore, has 8 possibilities. By the Multiplication Principle the number of such numbers is equal to

$$8 \cdot 8 = 64$$

Since the considered cases are mutually exclusive, then by the Addition Principle the total number of such numbers is equal to

$$72 + 64 + 64 = 200$$

and the right answer is $\boxed{\textbf{(C)} \ 200}$

Problem 2

Let z be the number of three-digit numbers that are divisible by 5 and their digits are all distinct. Find the product of the digits of z.

(A) 12   **(B)** 18   **(C)** 20   **(D)** 24   **(E)** 32

Solution

Notice that if the number is divisible by 5, then its units digit is either 5 or 0. Therefore, we have the numbers of the form $\overline{ab0}$ and $\overline{ab5}$. We will consider these two cases apart.

Case 1: the number is of the form $\overline{ab0}$. The digit a can take any value from 1 to 9 and, therefore, has 9 possibilities. The digit b cannot equal a nor 0 and, therefore, has 8 possibilities. By the Multiplication Principle the number of such numbers is equal to

$$9 \cdot 8 = 72$$

Case 2: the number is of the form $\overline{ab5}$. The digit a can take any value from 1 to 9 except 5 and, therefore, has 8 possibilities. The digit b cannot equal a nor 5 and, therefore, has 8 possibilities. By the Multiplication Principle the number of such numbers is equal to

$$8 \cdot 8 = 64$$

Since the considered cases are mutually exclusive, then by the Addition Principle the total number of such numbers is equal to

$$72 + 64 = 136$$

The product of the digits is equal $1 \cdot 3 \cdot 6 = 18$ and the right answer is $\boxed{\textbf{(B)} \ 18}$

Problem 3

How many numbers n less than 1000 and greater than 19 satisfy the following two properties: n is even and n gives remainder 1 when divided by 5?

(A) 81 **(B)** 89 **(C)** 90 **(D)** 91 **(E)** 98

Solution

Notice that if the number gives remainder 1 when divided by 5, then its units digit is either 1 or 6. Since n is even, then the last digit of n should be 6. There are two types of numbers less than 1000 and greater than 9: those that have two digits and those that have three digits. We will consider these two cases apart.

Case 1: n has two digits. Therefore, the number n is of the form $\overline{a6}$. The digit a can take any value from 2 to 9 and, therefore, has 8 possibilities and there are 8 such numbers.

Case 2: n has three digits. Therefore, the number n is of the form $\overline{ab6}$. The digit a can take any value from 1 to 9 and, therefore, has 9 possibilities. The digit b can take any value from 0 to 9 and, therefore, has 10 possibilities. There are $9 \cdot 10 = 90$ such numbers.

Since the considered cases are mutually exclusive, then by the Addition Principle the total number of such numbers is equal to

$$8 + 90 = 98$$

and the right answer is $\boxed{\textbf{(E)} \ 98}$

CHAPTER 44

PERMUTATIONS

Many problems require the counting of the possible **permutations** of the given objects. Let us assume that we are given n distinguishable objects. The number of possible permutations of these objects can be found as

$$n! = 1 \cdot 2 \cdot ... \cdot n$$

For example, if we are given three distinct numbers a, b, c and we need to find the total number of their permutations, then according to the above formula we have

$$3! = 1 \cdot 2 \cdot 3 = 6$$

Indeed, there are exactly 6 possible permutations of these numbers

$$(a, b, c)$$
$$(b, a, c)$$
$$(a, c, b)$$
$$(b, c, a)$$
$$(c, a, b)$$
$$(c, b, a)$$

Let us now consider several examples.

Problem 1

There are 7 horses in a race. It is known that the horse *Little Joe* always finishes first. In how many different orders can the horses finish the race?

(**A**) 12   (**B**) 24   (**C**) 120   (**D**) 720   (**E**) 5040

Solution

Notice that the first place if fixed and is taken by the horse *Little Joe*. The number of the remaining 6 positions can be filled with the other 6 horses in any order and, therefore, is equal to the number of permutations

$$6! = 1 \cdot 2 \cdot 3 \cdot 4 \cdot 5 \cdot 6 = 720$$

and the right answer is (**D**) 720

Problem 2

In how many ways can 3 boys and 4 girls be arranged on a bench if boys and girls alternate?

(**A**) 6   (**B**) 24   (**C**) 120   (**D**) 144   (**E**) 180

Solution

Notice that since there is one more girl than boy, then the positions of boys and girls are fixed

$$girl - boy - girl - boy - girl - boy - girl$$

where the girls occupy the positions 1, 3, 5 and 7, while boys occupy the positions 2, 4 and 6.

The number of ways to place the boys is equal to the number of permutations

$$3! = 1 \cdot 2 \cdot 3 = 6$$

The number of ways to place the girls is equal to the number of permutations

$$4! = 1 \cdot 2 \cdot 3 \cdot 4 = 24$$

Since these events are independent, then by the Multiplication Principle[1] the number of ways to arrange 3 boys and 4 girls is equal to

$$6 \cdot 24 = 144$$

[1] This principle is discussed in detail in Chapter 42 "Multiplication Principle"

and the right answer is $\boxed{\textbf{(D)} \ 144}$

Problem 3

Find the number of ways to place 4 physics books, 3 chemistry books and 2 mathematics books on a shelf, such that the books of the same subjects are placed together.

(A) 120 **(B)** 144 **(C)** 288 **(D)** 864 **(E)** 1728

Solution

Notice that the number of ways to permute the subjects is equal to

$$3! = 1 \cdot 2 \cdot 3 = 6$$

The number of ways to permute 4 physics is

$$4! = 1 \cdot 2 \cdot 3 \cdot 4 = 24$$

The number of ways to permute 3 chemistry books is

$$4! = 1 \cdot 2 \cdot 3 = 6$$

The number of ways to permute 2 mathematics books is

$$2! = 1 \cdot 2 = 2$$

Since these events are independent, then by the Multiplication Principle[2] the number of ways to arrange the books is equal to

$$6 \cdot 24 \cdot 6 \cdot 2 = 1728$$

and the right answer is $\boxed{\textbf{(E)} \ 1728}$

---

[2]This principle is discussed in detail in Chapter 42 "Multiplication Principle"

CHAPTER 45

COMBINATIONS

In many problems we need to know how to count the number of ways to choose m objects out of n given objects. This number is called a **combination** or "n choose m" and can be found as

$$\binom{n}{m} = \frac{n!}{m! \cdot (n-m)!}$$

For example, if we are given three distinct numbers a, b, c and we need to choose two of them, then according to the above formula we can do it in the following number of ways

$$\binom{3}{2} = \frac{3!}{2! \cdot (3-2)!} = \frac{6}{2 \cdot 1} = 3$$

Indeed, there are only 3 possible ways to choose 2 numbers

$$\{a, b\}, \{b, c\} \text{ and } \{a, c\}$$

Let us now consider several examples.

Problem 1

Given 10 points on a plane. How many ways are there to draw a segment?

(A) 45   **(B)** 55   **(C)** 100   **(D)** 120   **(E)** 144

Solution

Notice that in order to draw a segment we need to choose two distinct points from 10 given points. This can be done in the following number of ways

$$\binom{10}{2} = \frac{10!}{2! \cdot 8!} = 45$$

and the right answer is $\boxed{\textbf{(A)}\ 45}$

Problem 2

A committee of 4 people is to be chosen from a group of 5 men and 4 women. How many committees are possible if there are to be 2 men and 2 women?

(A) 45   **(B)** 56   **(C)** 60   **(D)** 72   **(E)** 84

Solution

Notice that in order to form a committee we need to choose 2 men out of 5 men and 2 women out of 4 women.

The number of ways to choose 2 men out of 5 men is equal to

$$\binom{5}{2} = \frac{5!}{2! \cdot 3!} = 10$$

The number of ways to choose 2 women out of 4 women is equal to

$$\binom{4}{2} = \frac{4!}{2! \cdot 4!} = 6$$

Since these events are independent, then by the Multiplication Principle[1] the number of ways to choose this committee is equal to

$$10 \cdot 6 = 60$$

and the right answer is $\boxed{\textbf{(C)}\ 60}$

[1]This principle is discussed in detail in Chapter 42 "Multiplication Principle"

Problem 3

How many ways are there to divide 10 boys into two basketball teams of 5 boys each?

(A) 120 **(B)** 126 **(C)** 144 **(D)** 252 **(E)** 288

Solution

Notice that the number of ways to choose the first team is equal to

$$\binom{10}{5} = \frac{10!}{5! \cdot 5!} = 252$$

The second team, therefore, is chosen automatically and consists of the boys who left. However, if instead we choose the first team to be the boys who left, then the other team is chosen automatically, which in fact represents the same choice of teams. Therefore, we overcount by the factor of 2 and the number of ways to choose two basketball teams of 5 boys each is equal to

$$\frac{252}{2} = 126$$

and the right answer is $\boxed{\textbf{(B)}\ 126}$

CHAPTER 46

COMPLEMENTARY COUNTING

Complementary Counting is a counting technique, where instead of counting the number of objects that satisfy certain condition, we count the total number of objects and then subtract the number of objects that do not satisfy the condition.

In the following discussion, if we are given a set A, then we will use the notation $|A|$ for the number of its elements.

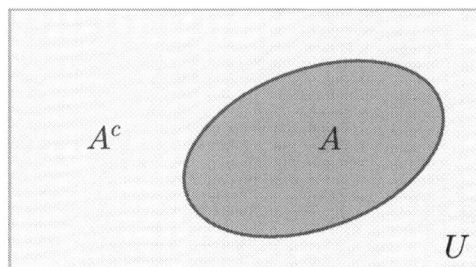

Let U be the set of all possible objects. Let A be the set of objects that satisfy the condition given in the problem and let A^c be the set of objects that do not satisfy this condition. The number of objects that satisfy the given condition can be found from the following equality

$$|A| = |U| - |A^c|$$

The set U is called the *universal set* and the set A^c is called the *complementary set* to A.

Let us now consider several examples where this technique can be applied.

Problem 1

There are four letters in some language. A word is any sequence of four letters, some two of which are the same. How many words are there in the language?

(A) 24   (B) 96   (C) 120   (D) 232   (E) 256

Solution

Let us count the total number of words and then subtract the number of words that have all different letters.

First, let us count the total number of words. Since each position can be filled with any letter, then the total number of words is equal to

$$4 \cdot 4 \cdot 4 \cdot 4 = 4^4 = 256$$

Now, let us count the number of words that have all different letters. It is equal to the number of permutations of four letters of the alphabet

$$4! = 1 \cdot 2 \cdot 3 \cdot 4 = 24$$

Therefore, the number of the words that have at least two letters that are the same is equal to

$$256 - 24 = 232$$

and the right answer is $\boxed{\textbf{(D) } 232}$

Problem 2

Let x be the number of five-digit numbers that have at least one even digit. What is the units digit of x?

(A) 0   (B) 1   (C) 2   (D) 3   (E) 5

Solution

Let us count the total number of five-digit numbers and then subtract the number of five-digit numbers that only have odd digits.

First, let us count the total number of five-digit numbers. Since the first digit cannot be zero, but all other digits can be any number from 0 to 9, then the total number of five-digit numbers is equal to

$$9 \cdot 10 \cdot 10 \cdot 10 \cdot 10 = 90000$$

Now, let us count the number of five-digit numbers that only have odd digits. Since every digit can be chosen only from 1, 3, 5, 7 or 9, then there are 5 options for each choice and the number of five-digit numbers that only have odd digits is equal to

$$5 \cdot 5 \cdot 5 \cdot 5 \cdot 5 = 5^5 = 3125$$

Therefore, the number of five-digit numbers that have at least one even digit is equal to

$$90000 - 3125 = 86875$$

and the right answer is $\boxed{\textbf{(E) } 5}$

Problem 3

Let A be the number of three-digit numbers that do not have any repeating 1s. What is the sum of digits of A?

(A) 12 **(B)** 15 **(C)** 16 **(D)** 18 **(E)** 20

Solution

Let us count the total number of three-digit numbers and then subtract the number of three-digit numbers that have at least two 1s.

First, let us count the total number of three-digit numbers. Since the first digit cannot be zero, but all other digits can be any number from 0 to 9, then the total number of three-digit numbers is equal to

$$9 \cdot 10 \cdot 10 = 900$$

Now, let us count the number of three-digit numbers that have at least two 1s. Let us consider the following cases: the number has exactly three 1s and the number has exactly two 1s.

If the number has exactly three 1s, then it is equal to 111.

If the number has exactly two 1s, then the numbers are of three types: $\overline{a11}$, $\overline{1a1}$ and $\overline{11a}$, where a is a digit distinct from 1. Since the first digit cannot be zero, then there are 8 numbers of the form $\overline{a11}$. For the numbers of the form $\overline{1a1}$ and $\overline{11a}$, the digit a can be any number from 0 to 9 except 1, and, therefore, there are 9 numbers of each type.

Therefore, the number of three-digit numbers that do not have any repeating 1s is equal to

$$900 - 1 - 8 - 9 - 9 = 873$$

which has the sum of its digits equal to $8 + 7 + 3 = 18$ and the right answer is $\boxed{\textbf{(D)} \ 18}$

CHAPTER 47

STARS AND BARS. PART 1

Stars and Bars is a method of solving the problem of distributing n indistinguishable stars into m distinct boxes by separating the stars using "bars". There are two separate cases when this method is applied, namely: *nonempty boxes* and *empty boxes*. In this chapter we will consider the case of the *nonempty boxes*.

For example, if we have 7 stars and we need to distribute them into 3 distinct nonempty boxes, we can do it with 2 bars located anywhere between the stars.

$$\bigstar \mid \bigstar \;\; \bigstar \mid \bigstar \;\; \bigstar \;\; \bigstar \;\; \bigstar$$

The number of possible distributions can be found in the following way. Let us assume that n stars are distributed into m distinct nonempty boxes. We can line up n stars in a row and then consider $n-1$ positions between the stars. A bar placed in such position represents the "separation" between the two consecutive boxes.

Since we only need $m-1$ bars, then the number of possible separations can be found as

$$\binom{n-1}{m-1} = \frac{(n-1)!}{(m-1)! \cdot (n-m)!}$$

Let us now consider several examples where this method can be applied.

Problem 1

How many ways are there to distribute 8 identical books to 5 girls if each girl should get a book?

(A) 10   **(B)** 35   **(C)** 40   **(D)** 56   **(E)** 70

Solution

Let us assume that the 8 books are the "stars". Notice that if we place the stars in a row, then there are 7 positions to place the 4 "bars" that will automatically distribute the stars into girls. Therefore, the number of possible distributions is equal to the number of ways to choose the 4 positions out of the possible 7 positions

$$\binom{7}{4} = \frac{7!}{4! \cdot 3!} = 35$$

and the right answer is $\boxed{\textbf{(B)}\ 35}$

Problem 2

How many ways are there to represent the number 10 as a sum of 4 positive integer numbers if the representations that differ in the order of the terms are different?

(A) 24   **(B)** 45   **(C)** 84   **(D)** 120   **(E)** 210

Solution

Let us assume that the 10 ones are the "stars"

$$1\ 1\ 1\ 1\ 1\ 1\ 1\ 1\ 1\ 1$$

Notice that if we place the stars in a row, then there are 9 positions to place the 3 "bars" that will automatically distribute the stars into numbers. Therefore, the number of ways to represent the number 10 as a sum of 4 positive integers is equal to the number of ways to choose the 3 positions out of the possible 9 positions

$$\binom{9}{3} = \frac{9!}{3! \cdot 6!} = 84$$

and the right answer is $\boxed{\textbf{(C)}\ 84}$

Problem 3

How many ways are there to put 5 white and 4 black cups in 3 boxes if all boxes are considered distinguishable and should contain at least one cup of each color?

(**A**) 20   (**B**) 28   (**C**) 24   (**D**) 12   (**E**) 18

Solution

Notice that the distributions of black and white cups are independent events. Therefore, we can consider them separately.

First, let us assume that the 5 white cups are the "stars". Notice that if we place the stars in a row, then there are 4 positions to place the 2 "bars" that will automatically distribute the cups into boxes. Therefore, the number of possible distributions is equal to the number of ways to choose the 2 positions out of the possible 4 positions

$$\binom{4}{2} = \frac{4!}{2! \cdot 2!} = 6$$

Now, let us assume that the 4 black cups are the "stars". Notice that if we place the stars in a row, then there are 3 positions to place the 2 "bars" that will automatically distribute the cups into boxes. Therefore, the number of possible distributions is equal to the number of ways to choose the 2 positions out of the possible 3 positions

$$\binom{3}{2} = \frac{3!}{2! \cdot 1!} = 3$$

Therefore, by the Multiplication Principle[1] the number of possible distributions is equal to

$$6 \cdot 3 = 18$$

and the right answer is $\boxed{(\textbf{E}) \ 18}$

---

[1] This principle is discussed in detail in Chapter 42 "Multiplication Principle"

CHAPTER 48

STARS AND BARS. PART 2

Stars and Bars is a method of solving the problem of distributing n indistinguishable stars into m distinct boxes by separating the stars using "bars". There are two separate cases when this method is applied, namely: *nonempty boxes* and *empty boxes*. In this chapter we will consider the case of the *empty boxes*.

For example, if we have 7 stars and we need to distribute them into 5 possibly empty boxes, we can do it with 4 bars permuted with the stars.

$$\bigstar \,\big|\, \bigstar \; \bigstar \,\big|\big|\, \bigstar \; \bigstar \; \bigstar \,\big|\, \bigstar$$

The number of possible distributions can be found in the following way. Let us assume that n indistinguishable stars are distributed into m distinct possibly empty boxes. We can line up n stars and $m-1$ bars in a row and then find the number of ways to choose $m-1$ positions for the bars among the $n+m-1$ possible positions.

This number can be found as

$$\binom{n+m-1}{m-1} = \frac{(n+m-1)!}{(m-1)! \cdot (n)!}$$

Let us now consider several examples where this method can be applied.

Problem 1

In how many ways can we distribute 15 identical apples to 4 distinct students if not all students have to get an apple?

(A) 60   **(B)** 330   **(C)** 816   **(D)** 1365   **(E)** 3060

Solution

Let us assume that the 15 apples are the "stars" and 4 students will be represented by 3 bars. Notice that if we place the stars and bars in a row, then there are $15 + 3 = 18$ objects. Therefore, the number of possible distributions is equal to

$$\binom{18}{3} = \frac{18!}{3! \cdot 15!} = 816$$

and the right answer is $\boxed{\textbf{(C)}\ 816}$

Problem 2

Find the number of nonnegative integer solutions of the equation

$$x_1 + x_2 + x_3 + x_4 + x_5 = 9$$

(A) 70   **(B)** 126   **(C)** 210   **(D)** 715   **(E)** 1001

Solution

Let us assume that the 9 ones are the "stars":

$$1\ 1\ 1\ 1\ 1\ 1\ 1\ 1\ 1$$

and the 5 variables are represented by 4 bars.

Notice that if we place the stars and bars in a row, then there are $9 + 4 = 13$ objects. Therefore, the number of possible distributions is equal to

$$\binom{13}{4} = \frac{13!}{4! \cdot 9!} = 715$$

and the right answer is $\boxed{\textbf{(D)}\ 715}$

Problem 3

How many ways are there to put 5 white and 4 black cups in 3 boxes if all boxes are considered distinguishable and might possibly be empty?[1]

(A) 315   **(B)** 84   **(C)** 40   **(D)** 10   **(E)** 4

Solution

Notice that the distributions of black and white cups are independent events. Therefore, we can consider them separately.

First, let us assume that the 5 white cups are the "stars" and the boxes are represented by 2 bars. Notice that if we place the stars and bars in a row, then there are $5 + 2 = 7$ objects. Therefore, the number of possible distributions is equal to

$$\binom{7}{2} = \frac{7!}{2! \cdot 5!} = 21$$

First, let us assume that the 4 black cups are the "stars" and the boxes are represented by 2 bars. Notice that if we place the stars and bars in a row, then there are $4 + 2 = 6$ objects. Therefore, the number of possible distributions is equal to

$$\binom{6}{2} = \frac{6!}{2! \cdot 4!} = 15$$

Therefore, by the Multiplication Principle[2] the number of possible distributions is equal to

$$21 \cdot 15 = 315$$

and the right answer is $\boxed{\textbf{(A)}\ 315}$

---

[1] A similar problem with nonempty boxes is solved in Chapter 47 "Stars and Bars. Part 1"
[2] This principle is discussed in detail in Chapter 42 "Multiplication Principle"

CHAPTER 49

INCLUSION-EXCLUSION PRINCIPLE. PART 1

Let us assume that we are given the sets X and Y. The set $X \cup Y$ represents the union of the sets X and Y and consists of all elements that belong to the set X or the set Y. The set $X \cap Y$ represents the intersection of the sets X and Y and consists of all elements that belong to both the set X and the set Y. The notation $|X|$ is used for the number of elements of the set X.

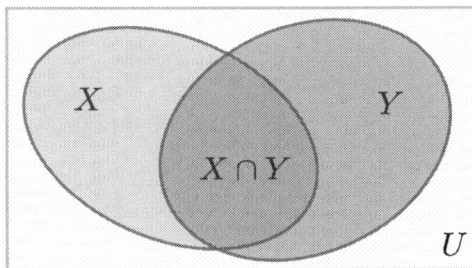

The **Principle of Inclusion and Exclusion** for two sets states that the following equality is true

$$|X \cup Y| = |X| + |Y| - |X \cap Y|$$

Let us now consider several examples, where we can use the Principle of Inclusion and Exclusion.

Problem 1

In some company there are 50 programmers: 25 are proficient in C, 30 are proficient in $Java$ and 10 in both. Determine the number of programmers who are proficient in C or $Java$.

(A) 5   **(B)** 15   **(C)** 45   **(D)** 50   **(E)** 55

Solution

Let X be the set of the programmers who are proficient in C and Y be the set of the programmers who are proficient in $Java$. Therefore, $X \cap Y$ is the set of the programmers that are proficient in both languages and $X \cup Y$ is the set of the programmers that are proficient in C or $Java$. Furthermore, we have

$$|X| = 25$$
$$|Y| = 30$$
$$|X \cap Y| = 10$$

Therefore, by the Principle of Inclusion and Exclusion

$$|X \cup Y| = |X| + |Y| - |X \cap Y| = 25 + 30 - 10 = 45$$

and the right answer is $\boxed{\textbf{(C)} \ 45}$

Problem 2

How many integers from 1 to 300 inclusively are multiples of 2 or 3?

(A) 50   **(B)** 100   **(C)** 150   **(D)** 200   **(E)** 250

Solution

Let X be the set of the integers from 1 to 300 that are multiples of 2 and Y be the set of the integers from 1 to 300 that are multiples of 3. Therefore, $X \cap Y$ is the set of the integers from 1 to 300 that are multiples of 2 and 3, i.e. are multiples of 6. Also $X \cup Y$ is the set of the integers from 1 to 300 that are multiples of 2 or 3. Thus we have

$$|X| = \frac{300}{2} = 150$$

$$|Y| = \frac{300}{3} = 100$$

$$|X \cap Y| = \frac{300}{6} = 50$$

Therefore, by the Principle of Inclusion and Exclusion

$$|X \cup Y| = |X| + |Y| - |X \cap Y| = 150 + 100 - 50 = 200$$

and the right answer is $\boxed{\textbf{(D)}\ 200}$

Problem 3

In some village there are 100 peasants and every peasant owns a cow, a goat or both. The number of peasants who own a cow is three times greater than the number of peasants who own a cow and a goat. How many peasants own only a cow if the number of peasants who only own a goat is 19?

(A) 54　　　**(B)** 27　　　**(C)** 18　　　**(D)** 81　　　**(E)** 46

Solution

Let X be the set of peasants who own a cow and Y be the set of peasants who own a goat. Therefore, $X \cap Y$ is the set of peasants who own a cow and a goat and $X \cup Y$ is the set of peasants who own a cow or a goat. Let a be the number of peasants who own a cow and a goat. Therefore

$$|X \cap Y| = a$$
$$|X| = 3a$$
$$|Y| = a + 19$$

The Principle of Inclusion and Exclusion

$$|X \cup Y| = |X| + |Y| - |X \cap Y|$$

becomes an equation

$$(3a) + (a + 19) - (a) = 100$$

We can solve this equation by isolating a

$$3a + a + 19 - a = 100$$
$$3a + 19 = 100$$

$$3a = 81$$
$$a = 27$$

The number of peasants who own only a cow is equal to

$$3a - a = 2a = 2 \cdot 27 = 54$$

and the right answer is $\boxed{\textbf{(A)}\ 54}$

CHAPTER 50

INCLUSION-EXCLUSION PRINCIPLE. PART 2

Let us assume that we are given the sets X, Y and Z. Let us assume the following

- $X \cup Y \cup Z$ represents the union of the sets X, Y and Z and consists of all elements that belong to the set X, the set Y or the set Z

- The set $X \cap Y$ represents the intersection of the sets X and Y and consists of all elements that belong to both the set X and the set Y

- The set $Y \cap Z$ represents the intersection of the sets Y and Z and consists of all elements that belong to both the set Y and the set Z

- The set $Z \cap X$ represents the intersection of the sets Z and X and consists of all elements that belong to both the set Z and the set X

- The notations $|X|$, $|Y|$ and $|Z|$ are used for the number of elements of the sets X, Y and Z respectively.

The **Principle of Inclusion and Exclusion** for three sets states that the following equality is true

$$|X \cup Y \cup Z| = |X| + |Y| + |Z| - |X \cap Y| - |Y \cap Z| - |Z \cap X| + |X \cap Y \cap Z|$$

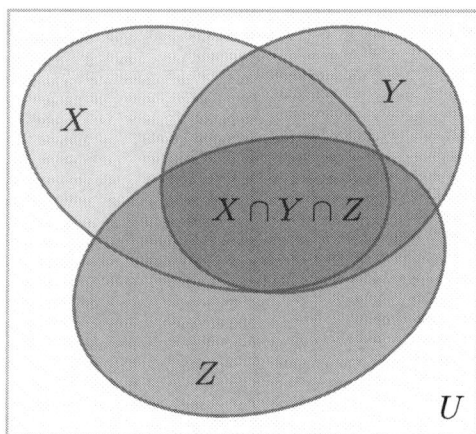

Let us now consider several examples, where we can use the Principle of Inclusion and Exclusion.

Problem 1

There are 200 senior students in some school. Each senior student takes Mathematics, History or Art, but no student takes all three subjects at the same time. It is known that the number of students who take any two of these subjects is the same. Which of the following represents the number of students that take Mathematics if 95 students take History and 111 students take Art?

(A) 55   **(B)** 56   **(C)** 68   **(D)** 71   **(E)** 75

Solution

Let X, Y and Z be the sets of the students who take Mathematics, History and Art respectively. Let x be the number of students who take Mathematics. Therefore, we have

$$|X \cup Y \cup Z| = 200$$
$$|X \cap Y \cap Z| = 0$$
$$|X| = x$$
$$|Y| = 95$$
$$|Z| = 111$$

Let k be the number of students who take any two of the subjects. Then we have

$$|X \cap Y| = k$$
$$|Y \cap Z| = k$$
$$|Z \cap X| = k$$

Therefore, from the Principle of Inclusion and Exclusion

$$|X \cup Y \cup Z| = |X| + |Y| + |Z| - |X \cap Y| - |Y \cap Z| - |Z \cap X| + |X \cap Y \cap Z|$$

we have the following equation

$$200 = x + 95 + 111 - k - k - k + 0$$
$$200 = x + 206 - 3k$$
$$3k - 6 = x$$

From here we see that the number x is a multiple[1] of 3. The only number in the list of answers that is a multiple of 3 is 75 and the right answer is $\boxed{\textbf{(E) } 75}$

Problem 2

How many integers from 1 to 300 inclusively are multiples of 2, 3 or 5?

(A) 70   **(B)** 80   **(C)** 100   **(D)** 150   **(E)** 220

Solution

Let X be the set of integers from 1 to 300 that are multiples of 2, Y be the set of integers from 1 to 300 that are multiples of 3 and Z be the set of integers from 1 to 300 that are multiples of 5. Therefore, we have

$$|X| = \frac{300}{2} = 150$$
$$|Y| = \frac{300}{3} = 100$$
$$|Y| = \frac{300}{5} = 60$$

Also $X \cap Y$ is the set of the integers from 1 to 300 that are multiples of 2 and 3, i.e. are multiples of 6, $Y \cap Z$ is the set of the integers from 1 to 300 that are multiples of 3 and 5, i.e. are multiples of 15 and $Z \cap X$ is the set of the integers from 1 to 300

---

[1] You can find more problems that use similar arguments in Chapter 36 "Divisibility"

that are multiples of 5 and 2, i.e. are multiples of 10. Therefore, we have

$$|X \cap Y| = \frac{300}{6} = 50$$
$$|Y \cap Z| = \frac{300}{15} = 20$$
$$|Z \cap X| = \frac{300}{10} = 30$$

The set $X \cap Y \cap Z$ is the set of the integers from 1 to 300 that are multiples of 2, 3 and 5, i.e. are multiples of 30. Therefore

$$|X \cap Y \cap Z| = \frac{300}{30} = 10$$

The set $X \cup Y \cup Z$ is the set of the integers from 1 to 300 that are multiples of 2, 3 or 5.

Therefore, from the Principle of Inclusion and Exclusion

$$|X \cup Y \cup Z| = |X| + |Y| + |Z| - |X \cap Y| - |Y \cap Z| - |Z \cap X| + |X \cap Y \cap Z|$$

we have

$$|X \cup Y \cup Z| = 150 + 100 + 60 - 50 - 20 - 30 + 10 = 220$$

and the right answer is $\boxed{\textbf{(E)}\ 220}$

Problem 3

In some village there are 113 peasants and each peasant owns a certain number of cows, goats and chickens. The number of peasants who own chickens is eight times greater than the number of peasants who own cows and goats. The number of peasants who own cows is equal to the number of peasants who own goats and chickens, and the number of peasants who own goats is equal to the number of peasants who own cows and chickens. Which of the following intervals does not contain a possible number of peasants who own cows, goats and chickens at the same time?

(A) $[33, 38]$ **(B)** $[40, 45]$ **(C)** $[52, 55]$ **(D)** $[56, 60]$ **(E)** $[62, 67]$

Solution

Let X be the set of peasants who own cows, Y be the set of peasants who own goats and Z be the set of peasants who own chickens. Let x be the number of peasants who own cows, goats and chickens. Therefore

$$|X \cup Y \cup Z| = 113$$
$$|X \cap Y \cap Z| = x$$

We also have that $X \cap Y$ is the set of peasants who own cows and goats, $Y \cap Z$ is the set of peasants who own goats and chickens, and $Z \cap X$ is the set of peasants who own chickens and cows. Let us assume that

$$|X \cap Y| = a$$
$$|Y \cap Z| = b$$
$$|Z \cap X| = c$$

By the conditions given in the problem we have

$$|Z| = 8a$$
$$|X| = b$$
$$|Y| = c$$

From the Principle of Inclusion and Exclusion

$$|X \cup Y \cup Z| = |X| + |Y| + |Z| - |X \cap Y| - |Y \cap Z| - |Z \cap X| + |X \cap Y \cap Z|$$

we have

$$113 = 8a + b + c - a - b - c + x$$
$$113 = 7a + x$$
$$113 - 7a = x$$

Notice that from here

$$x = 113 - 7a \equiv 1 \pmod{7}$$

The only interval that does not contain a number that gives remainder[2] 1 modulo 7 is $[52, 55]$ and the right answer is $\boxed{\textbf{(C)} \; [52, 55]}$

[2] You can find more problems that use similar arguments in Chapter 37 "Remainders"

CHAPTER 51

INSCRIBED ANGLES

Let A and B be two points on the circle with center O. The angle $\angle AOB$ is called **central angle**. If the points C, D and E also lie on the circle with center O, then each of the angles $\angle ACB$, $\angle ADB$ and $\angle AEB$ is called **inscribed angle**.

If the points A and B form the diameter of the circle, then the inscribed angle $\angle ACB$ is right

$$\angle ACB = 90°$$

If the points C and O lie in the same semiplane with respect to the line AB, then the measure of the central angle $\angle AOB$ is twice greater than the measure of the inscribed angle $\angle ACB$

$$\angle AOB = 2 \cdot \angle ACB$$

If the points C and D lie in the same semiplane with respect to the line AB, then the measures of the inscribed angles $\angle ACB$ and $\angle ADB$ are the same

$$\angle ACB = \angle ADB$$

If the points C and E lie in different semiplanes with respect to the line AB, then the inscribed angles $\angle ACB$ and $\angle AEB$ are supplementary

$$\angle ACB + \angle AEB = 180°$$

If the points O and E lie in different semiplanes with respect to the line AB, then the measure of the arc $\overset{\frown}{AEB}$ is equal to the measure of the central angle $\angle AOB$

$$\overset{\frown}{AEB} = \angle AOB$$

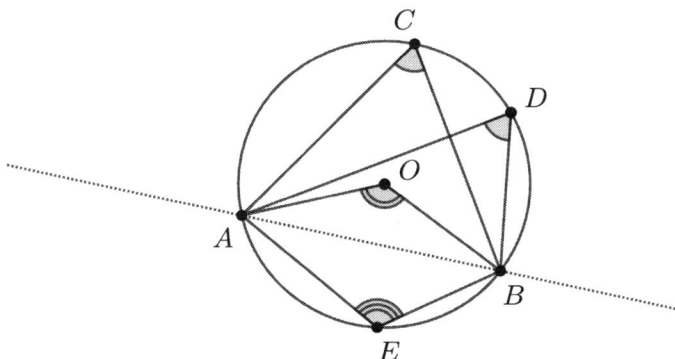

Problem 1

Let O be the circumcenter of the obtuse triangle ABC ($\angle ACB > 90°$). If $\angle AOB = x$ and $\angle ACB = y$, which of the following is true?

(A) $x = 90°$ **(B)** $y = 2x$ **(C)** $x = 2y$ **(D)** $x + y = 180°$ **(E)** $x + 2y = 360°$

Solution

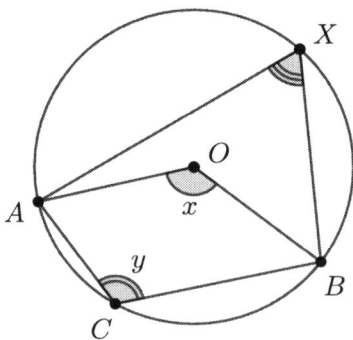

Since $\angle ACB > 90°$, then the points O and C are located in different semiplanes with respect to the line AB. Let X be any point on the circumcircle and in the same semiplane as the point O with respect to the line AB. Therefore

$$\angle AOB = 2 \cdot \angle AXB$$
$$x = 2 \cdot \angle AXB$$
$$\frac{x}{2} = \angle AXB$$

Since the inscribed angles $\angle AXB$ and $\angle ACB$ add up to $180°$ we have

$$\angle ACB + \angle AEB = 180°$$
$$y + \frac{x}{2} = 180°$$
$$x + 2y = 360°$$

and the right answer is $\boxed{\textbf{(E) } x + 2y = 360°}$

Problem 2

The quadrilateral $ABCD$ is inscribed into a circle with center O, such that AD is its diameter. If the ratio of the measures of the angles $\angle ACB$ and $\angle BAD$ is $2 : 7$, which of the following is the measure of the angle $\angle BCD$?

(A) $90°$ **(B)** $100°$ **(C)** $110°$ **(D)** $120°$ **(E)** $150°$

Solution

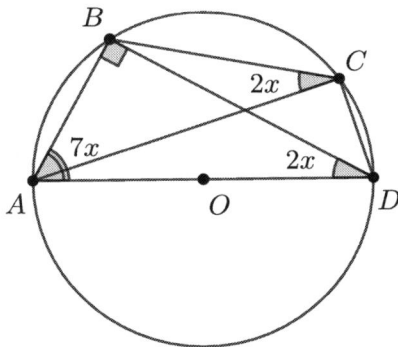

Let $\angle ACB = 2x$ and $\angle BAD = 7x$. Therefore, we have

$$\angle ADB = \angle ACB = 2x$$

Since AD is the diameter, then $\angle ABD = 90°$. From the triangle ABD we have

$$\angle ABD + \angle BAD + \angle ADB = 180°$$
$$90° + 7x + 2x = 180°$$
$$90° + 9x = 180°$$
$$9x = 90°$$
$$x = 10°$$

From here we have

$$\angle BAD = 7x = 7 \cdot 10° = 70°$$

Since the inscribed angles $\angle BAD$ and $\angle BCD$ add up to $180°$, we have

$$\angle BAD + \angle BCD = 180°$$
$$70° + \angle BCD = 180°$$
$$\angle BCD = 110°$$

and the right answer is $\boxed{\textbf{(C)}\ 110°}$

Problem 3

$ABCDE$ is a cyclic pentagon, such that $BD \parallel AE$. It is known that the measure of the angle $\angle ACB$ in degrees represents a positive integer. Which of the following represents the possible value of the angle $\angle BED$ if $\angle ABE = 5\angle AEB$?

(A) $66°$ **(B)** $67°$ **(C)** $68°$ **(D)** $69°$ **(E)** $70°$

Solution

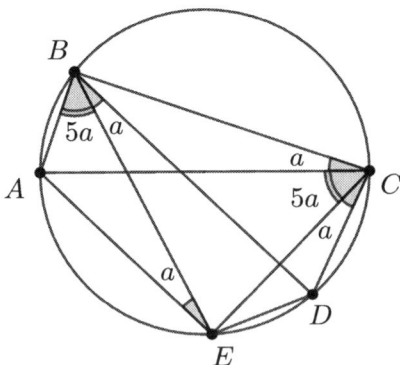

Let us put $\angle AEB = a$. Then we have

$$\angle ABE = 5\angle AEB = 5a$$
$$\angle ACB = \angle AEB = a$$
$$\angle EBD = \angle AEB = a$$
$$\angle ECD = \angle EBD = a$$
$$\angle ACE = \angle ABE = 5a$$

From here

$$\angle BCD = \angle ACB + \angle ACE + \angle ECD = a + a + 5a = 7a$$

and

$$\angle BED = 180° - \angle BCD = 180 - 7a$$

We can see that the measure of the angle $\angle BED$ is congruent to 5 modulo 7. The only angle in the list that is congruent[1] to 5 modulo 7 is $68°$ and, therefore, the right answer is $\boxed{\textbf{(C) } 68°}$

---

[1] You can find more problems that use similar arguments in Chapter 37 "Remainders"

CHAPTER 52

PYTHAGOREAN THEOREM

Let us assume that we are given a right triangle ABC with the angle $\angle ACB = 90°$ and the sides $AB = c$, $BC = a$ and $AC = b$.

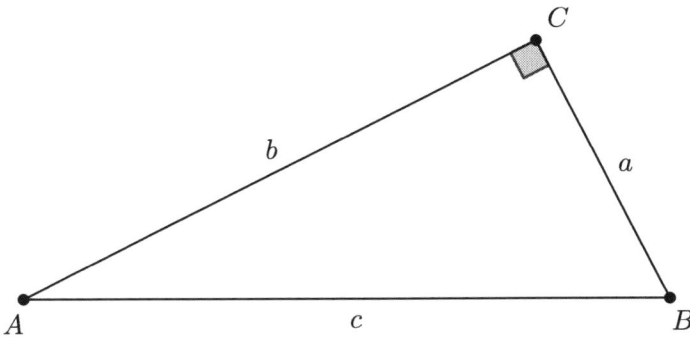

The **Pythagorean Theorem** states that the sides of the right triangle ABC satisfy the following equality

$$a^2 + b^2 = c^2$$

Notice that if we are given any two sides of a right triangle, then from this equality we can easily find the third side. This makes the Pythagorean Theorem extremely useful for the problems where we are given a right triangle with two known sides.

Let us consider several examples.

Problem 1

Find the area of the triangle with the sides 7, 7 and 8.

(A) $\sqrt{33}$ **(B)** $2\sqrt{33}$ **(C)** $4\sqrt{11}$ **(D)** $6\sqrt{11}$ **(E)** $4\sqrt{33}$

Solution

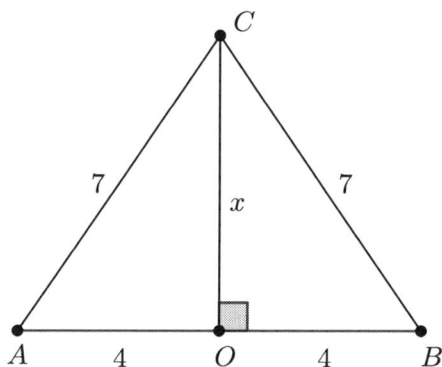

Let us draw the altitude to the side of length 8. Let the length of the altitude be x. Notice that the original triangle is isosceles and, therefore, the altitude is also a median. Therefore, we have two right triangles with hypotenuse 7 and one of the legs 4. From the Pythagorean Theorem we can find the length of the altitude

$$a^2 + b^2 = c^2$$
$$(x)^2 + (4)^2 = (7)^2$$
$$x^2 + 16 = 49$$
$$x^2 = 33$$
$$x = \sqrt{33}$$

The area S of the original triangle can be found using its base and height

$$S = \frac{b \cdot h}{2} = \frac{8 \cdot \sqrt{33}}{2} = 4\sqrt{33}$$

and the right answer is $\boxed{\textbf{(E)}\ 4\sqrt{33}}$ [1]

Problem 2

Given a semicircle of radius R and a rectangle 4×3 inscribed into the semicircle, such that the side of length 4 lies on the diameter of the semicircle. Let the length of the radius of the semicircle be equal $\sqrt{n}$, where n is a positive integer. Which of the following represents the sum of digits of the number n?

(A) 4   **(B)** 5   **(C)** 8   **(D)** 10   **(E)** 11

Solution

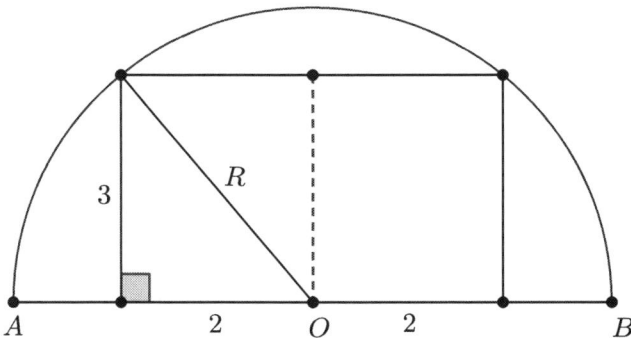

Notice that the center of the semicircle is the midpoint of the side of the rectangle of length 4. Therefore, we have a right triangle with the hypotenuse R and the legs 3 and 2. Now we can find the length of the hypotenuse R from the Pythagorean Theorem

$$a^2 + b^2 = c^2$$
$$(3)^2 + (2)^2 = (R)^2$$
$$9 + 4 = R^2$$
$$13 = R^2$$
$$\sqrt{13} = R$$

From here we have $n = 13$. The sum of digits of n is equal to $1 + 3 = 4$ and the right answer is $\boxed{\textbf{(A)}\ 4}$

---

[1] This problem can be also solved using the Heron's Formula discussed in detail in Chapter 55 "Heron's Formula"

Problem 3

The perimeter of the rhombus $ABCD$ is $30\sqrt{10}$ units. One of its diagonals is 12 units longer that the other. Which of the following is closest to the value of the area of the rhombus?

(A) 515   **(B)** 519   **(C)** 524   **(D)** 527   **(E)** 531

Solution

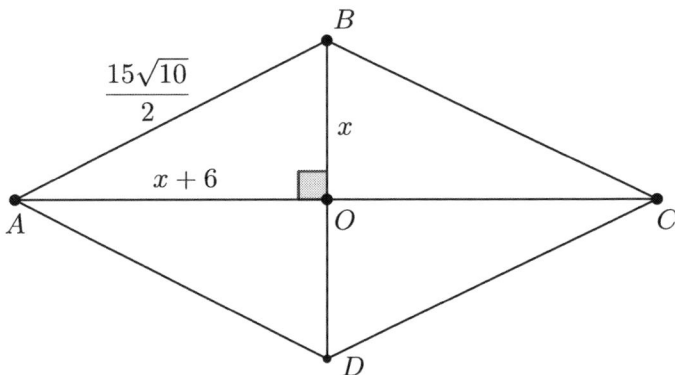

Notice that since all the sides of the rhombus are equal then the length of each side is equal to

$$\frac{30\sqrt{10}}{4} = \frac{15\sqrt{10}}{2}$$

Let O be the point of intersection of the diagonals of the rhombus. Since the diagonals of any rhombus are perpendicular and bisect each other, then the triangle ABO is a right triangle with hypotenuse $AB = \frac{15\sqrt{10}}{2}$. If the leg $BO = x$, then the leg $AO = x + 6$ and by the Pythagorean Theorem we have

$$a^2 + b^2 = c^2$$

$$(x)^2 + (x+6)^2 = \left(\frac{15\sqrt{10}}{2}\right)^2$$

$$2x^2 + 12x + 36 = \frac{1125}{2}$$

$$4x^2 + 24x + 72 = 1125$$

$$4x^2 + 24x - 1053 = 0$$

This equation is quadratic and can be solved using the Quadratic Formula[2]

$$x = \frac{-(24) \pm \sqrt{(24)^2 - 4(4)(-1053)}}{2(4)} = \frac{-24 \pm \sqrt{17424}}{8} = \frac{-24 \pm 132}{8}$$

The positive solution of this equation is

$$x = \frac{-24 + 132}{8} = \frac{27}{2} = 13.5$$

Therefore, $BO = 13$, $AO = 13.5 + 6 = 19.5$ and the area of the triangle ABO is equal to

$$\frac{13.5 \cdot 19.5}{2} = 131.625$$

Since the rhombus $ABCD$ consists of four congruent triangles of area 131.625, then the area of the rhombus is

$$4 \cdot 131.625 = 526.5$$

the right answer is $\boxed{\textbf{(D)} \ 527}$

---

[2]This formula is discussed in detail in Chapter 13 "Quadratic Formula"

CHAPTER 53

SHOELACE FORMULA

Let us assume that we are given a polygon on the coordinate plane that has n vertices defined by the coordinates:

$$(x_1, y_1)$$

$$(x_2, y_2)$$

$$(x_3, y_3)$$

$$\ldots$$

$$(x_{n-1}, y_{n-1})$$

$$(x_n, y_n)$$

The **Shoelace Formula** states that the area S of the polygon can be found using the following formula

$$S = \frac{1}{2} |x_1 y_2 + x_2 y_3 + \ldots + x_n y_1 - y_1 x_2 - y_2 x_3 - \ldots - y_n x_1|$$

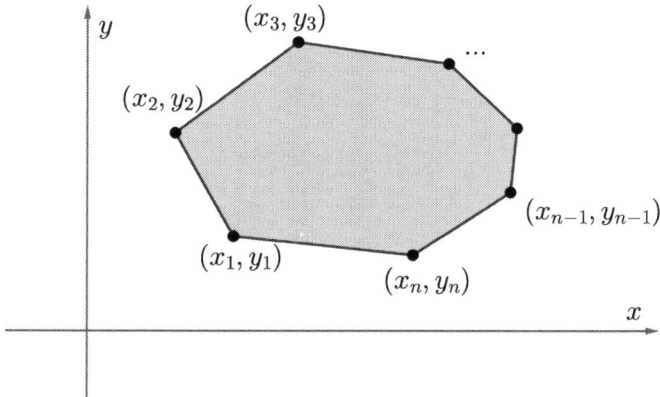

Notice that the formula works for any polygon, including triangles and quadrilaterals. Shoelace Formula is very useful for the problems where we need to find the area of a polygon given the coordinates of its vertices.

Let us consider several examples.

Problem 1

Which of the following is the area of the quadrilateral formed by the vertices $(0,0)$, $(4,1)$, $(5,3)$, $(-1,5)$?

(A) 15.5   **(B)** 16.0   **(C)** 16.5   **(D)** 17.0   **(E)** 17.5

Solution

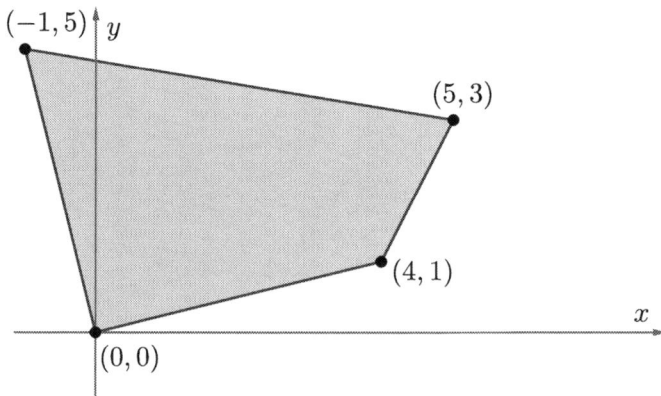

Notice that the quadrilateral is formed by the following vertices

$$(x_1, y_1) = (0, 0)$$
$$(x_2, y_2) = (4, 1)$$
$$(x_3, y_3) = (5, 3)$$
$$(x_4, y_4) = (-1, 5)$$

Therefore, by the Shoelace Formula the area of the triangle is equal to

$$\frac{1}{2} |x_1 y_2 + x_2 y_3 + x_3 y_4 + x_4 y_1 - y_1 x_2 - y_2 x_3 - y_3 x_4 - y_4 x_1| =$$

$$\frac{1}{2} |0 \cdot 1 + 4 \cdot 3 + 5 \cdot 5 + (-1) \cdot 0 - 0 \cdot 4 - 1 \cdot 5 - 3 \cdot (-1) - 5 \cdot 0| = 17.5$$

and the right answer is (E) 17.5

Problem 2

A parallelogram with vertices at the points $(0,0)$, $(1,4)$, $(7,4)$ and $(6,0)$ contains the triangle with vertices at the points $(2,1)$, $(3,3)$, and $(6,2)$. What is the area of the part of the parallelogram that is not covered by the triangle?

(A) 20.0   (B) 20.5   (C) 21.0   (D) 21.5   (E) 22.0

Solution

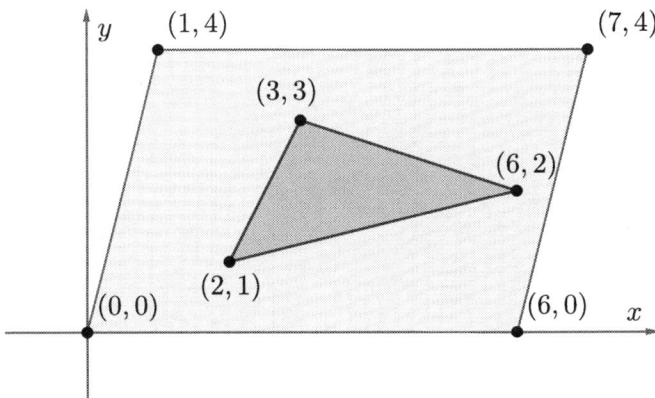

Notice that the area of the parallelogram can be found by multiplying its base by its height

$$6 \cdot 4 = 24$$

Let the vertices of the triangle be

$$(x_1, y_1) = (2, 1)$$
$$(x_2, y_2) = (3, 3)$$
$$(x_3, y_3) = (6, 2)$$

Therefore, by the Shoelace Formula the area of the triangle is equal to

$$\frac{1}{2}|x_1y_2 + x_2y_3 + x_3y_1 - y_1x_2 - y_2x_3 - y_3x_1| =$$
$$\frac{1}{2}|2 \cdot 3 + 3 \cdot 2 + 6 \cdot 1 - 1 \cdot 3 - 3 \cdot 6 - 2 \cdot 2| = 3.5$$

The area of the parallelogram that is not covered by the triangle is

$$24 - 3.5 = 20.5$$

and the right answer is $\boxed{\textbf{(B) } 20.5}$

Problem 3

Given a positive real number a and the triangle with the vertices at the points $(1, 1)$, (a, a^2) and $(a^2 - 1, a + 2)$. Which of the following is closest to the value of a if the area of the triangle is equal to 11.

(A) 2.6   **(B)** 2.8   **(C)** 3.0   **(D)** 3.2   **(E)** 3.4

Solution

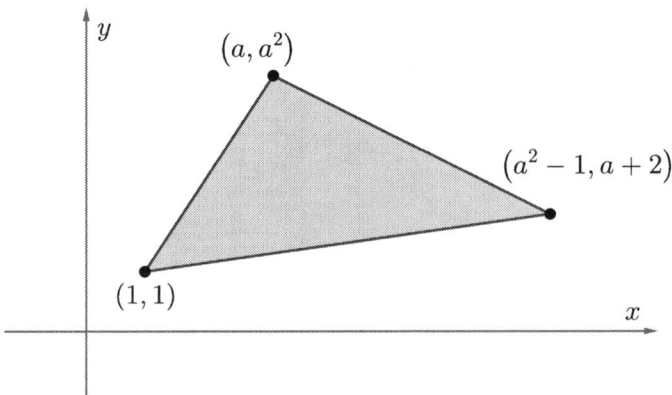

Let the given vertices be

$$(x_1, y_1) = (1, 1)$$
$$(x_2, y_2) = (a, a^2)$$
$$(x_3, y_3) = (a^2 - 1, a + 2)$$

Therefore, by the Shoelace Formula the area of the triangle is equal to

$$\frac{1}{2} \left| x_1 y_2 + x_2 y_3 + x_3 y_1 - y_1 x_2 - y_2 x_3 - y_3 x_1 \right|$$

By substituting the coordinates we have

$$\frac{1}{2} \left| 1 \cdot a^2 + a \cdot (a + 2) + (a^2 - 1) \cdot 1 - 1 \cdot a - a^2 \cdot (a^2 - 1) - (a + 2) \cdot 1 \right|$$

which is equivalent to

$$\frac{1}{2} \left| a^4 - 4a^2 + 3 \right|$$

Since the area equals to 11, then we have the following equation

$$\frac{1}{2} \left| a^4 - 4a^2 + 3 \right| = 11$$
$$\left| a^4 - 4a^2 + 3 \right| = 22$$
$$a^4 - 4a^2 + 3 = \pm 22$$
$$a^4 - 4a^2 + 3 \mp 22 = 0$$

Let us make a substitution $u = a^2$, where $u \geq 0$. Now we have two quadratic equations

$$u^2 - 4u + 25 = 0$$

and

$$u^2 - 4u - 19 = 0$$

The first equation has no real solutions, while second equation has a positive solution that can be found using the Quadratic Formula[1]

$$u = \frac{-(-4) \pm \sqrt{(-4)^2 - 4(1)(-19)}}{2(1)}$$
$$= \frac{4 \pm \sqrt{92}}{2}$$
$$= \frac{4 \pm 2\sqrt{23}}{2}$$
$$= 2 \pm \sqrt{23}$$

---

[1] This formula is discussed in detail in Chapter 13 "Quadratic Formula"

The positive solution is $u = 2 + \sqrt{23}$ and from here

$$a = \sqrt{2 + \sqrt{23}}$$

Notice that we can estimate[2] the value of a as follows

$$\sqrt{2 + \sqrt{23}} < \sqrt{2 + \sqrt{25}} = \sqrt{2 + 5} = \sqrt{7} = \sqrt{\frac{700}{100}} < \sqrt{\frac{729}{100}} = \frac{27}{10} = 2.7$$

and, therefore, the right answer is $\boxed{\textbf{(A)}\ 2.6}$

---

[2] You can find more problems that use these techniques in Chapter 25 "Estimations"

CHAPTER 54

AREA OF A TRIANGLE

Let us assume that we are given a triangle ABC. Let the side $AC = b$ be its base and $BD = h$ be its height drawn from the vertex B.

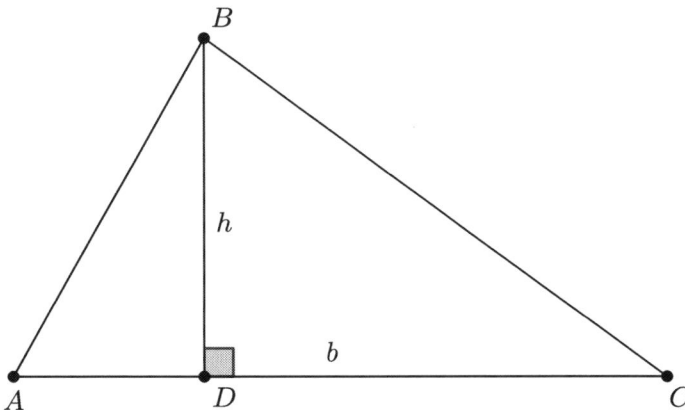

The area S of the triangle ABC can be found as

$$S = \frac{b \cdot h}{2}$$

This is one of the most frequently used formulas for the **area of a triangle**. Let us consider several examples of the application of this formula.

Problem 1

The point M is chosen on the side AC of the triangle ABC, such that $AM : MC = 2 : 3$. What is the ratio of the area of the triangle AMB to the triangle CMB?

(A) $2 : 3$ **(B)** $3 : 2$ **(C)** $4 : 9$ **(D)** $9 : 4$ **(E)** $1 : 1$

Solution

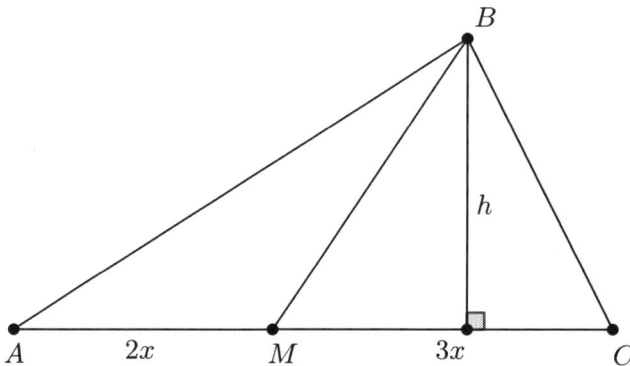

Notice that the triangles AMB and CMB share the same height, let us call it h. Let $AM = 2x$ and $MC = 3x$. Then the areas of the triangles AMB and CMB are

$$S_{AMB} = \frac{2x \cdot h}{2}$$

$$S_{CMB} = \frac{3x \cdot h}{2}$$

The ratio of the areas, therefore, is equal to

$$\frac{S_{AMB}}{S_{CMB}} = \frac{\frac{2xh}{2}}{\frac{3xh}{2}} = \frac{2xh}{2} \cdot \frac{2}{3xh} = \frac{2\cancel{x}\cancel{h}}{\cancel{2}} \cdot \frac{\cancel{2}}{3\cancel{x}\cancel{h}} = \frac{2}{3}$$

and the right answer is $\boxed{\textbf{(A)}\ 2 : 3}$

Problem 2

Given the triangle ABC with angles $A = 45°$, $B = 45°$ $C = 90°$. A semicircle with center O is inscribed into the triangle, such that its diameter lies on the larger side of the triangle. Find the area of the triangle if the area of the semicircle is 50π.

(A) 100   **(B)** 180   **(C)** 200   **(D)** 240   **(E)** 320

Solution

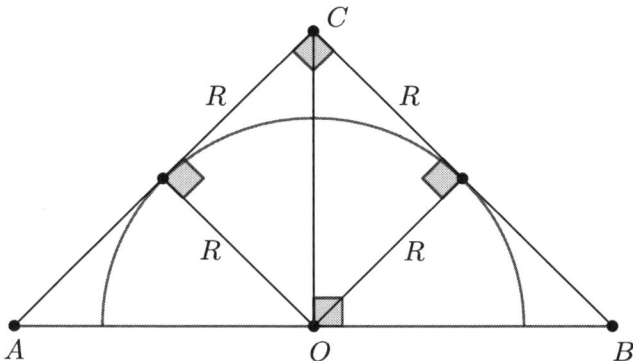

Let us start by finding the radius of the circle. Let R be the radius of the semicircle. Since the area of the complete circle is given by the formula πR^2 and half the area is equal to 50π, then

$$50\pi = \frac{1}{2}\pi R^2$$

We can solve this equation by isolating R

$$50\pi = \frac{1}{2}\pi R^2$$
$$100\pi = \pi R^2$$
$$100\not\pi = \not\pi R^2$$
$$100 = R^2$$
$$10 = R$$

Notice that the triangle ABC is an isosceles right triangle and the diameter of the semicircle lies on its hypotenuse. Let us draw the radii to the points of tangency of the circle with the legs of the triangle. Since they are perpendicular to the legs, then the formed quadrilateral is a square of side 10. Therefore, the triangle ABC is divided into 4 congruent isosceles right triangles. The legs of the triangles equal 10 and, therefore, the area of the triangle ABC can be found as

$$4 \cdot \frac{10 \cdot 10}{2} = 200$$

and the right answer is $\boxed{\textbf{(C) } 200}$

Problem 3

The area of the triangle formed by the lines $y = x$, $y = a$, $y = a^2x$ is equal to 180. Which of the following is closest to the value of a?

(A) 18.8   **(B)** 19.5   **(C)** 20.3   **(D)** 21.0   **(E)** 21.3

Solution

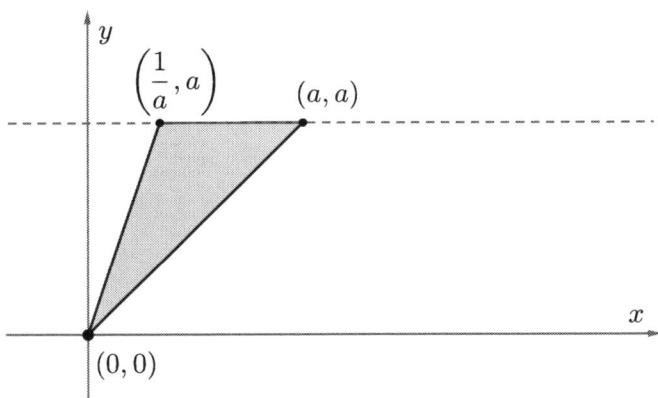

Let us find the coordinates of the vertices of the triangle. The lines $y = x$ and $y = a^2x$ both pass through the origin, therefore, $(0,0)$ is one of the vertices of the triangle[1]. Another vertex is (a, a), which lies on both lines $y = x$ and $y = a$. The y-coordinate of the third vertex is a and the x-coordinate can be found from the equation

$$a^2x = a$$
$$x = \frac{a}{a^2}$$
$$x = \frac{1}{a}$$

Therefore, the third vertex is $\left(\frac{1}{a}, a\right)$.

---

[1] You can find more similar problems in Chapter 30 "Lines on a Coordinate Plane"

The base of the triangle is equal to $\left|a - \frac{1}{a}\right|$, while the height of the triangle is equal to a. Since the area of the triangle is 180, then we have the following equation

$$\frac{\left|a - \frac{1}{a}\right| \cdot a}{2} = 180$$

$$\frac{\left|a^2 - 1\right|}{2} = 180$$

$$\left|a^2 - 1\right| = 360$$

$$a^2 - 1 = \pm 360$$

$$a^2 = \pm 360 + 1$$

From here we have $a = 19$ and the right answer is $\boxed{\textbf{(A) } 18.8}$ [2]

---

[2]This problem can be also solved using the formula discussed in detail in Chapter 53 "Shoelace Formula"

CHAPTER 55

HERON'S FORMULA

Let us assume that we are given a triangle ABC with the sides of length a, b and c and let p be its semiperimeter defined as

$$p = \frac{a + b + c}{2}$$

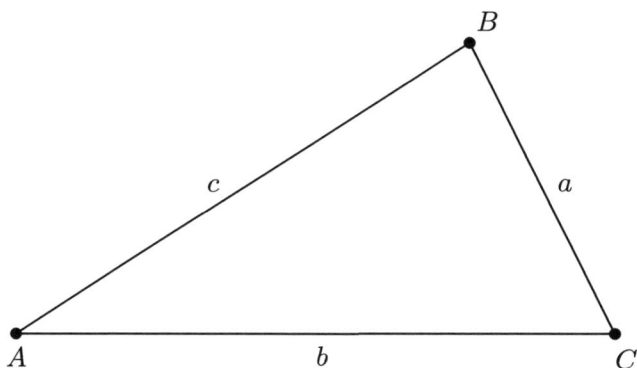

The **Heron's Formula** states that the area S of the triangle ABC can be found by the following formula

$$S = \sqrt{p \cdot (p - a) \cdot (p - b) \cdot (p - c)}$$

The Heron's Formula is very useful for the problems where we need to find the area of the triangle when the lengths of all its three sides are given. Let us consider several examples.

Problem 1

Find the area of the triangle with the sides 9, 10 and 11.

(A) $30\sqrt{2}$ **(B)** $31\sqrt{2}$ **(C)** $31\sqrt{3}$ **(D)** $32\sqrt{5}$ **(E)** $30\sqrt{5}$

Solution

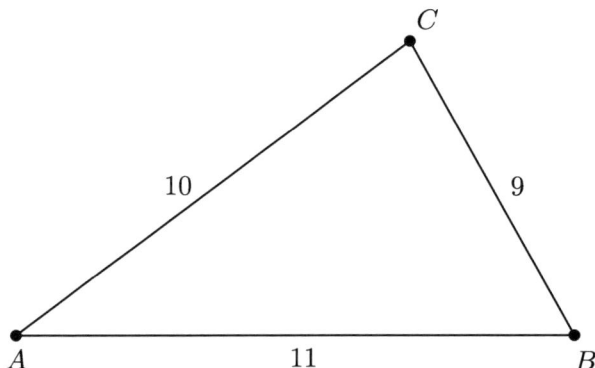

Notice that since all three sides of the triangle are given, then we can use the Heron's formula. Let $a = 9$, $b = 10$ and $c = 11$. First, let us find the semiperimeter p of the triangle

$$p = \frac{9 + 10 + 11}{2} = \frac{30}{2} = 15$$

The values of the expressions $p - a$, $p - b$ and $p - c$ are

$$p - a = 15 - 9 = 6$$
$$p - b = 15 - 10 = 5$$
$$p - c = 15 - 11 = 4$$

By Heron's formula the area of the triangle is equal to

$$S = \sqrt{15 \cdot 6 \cdot 5 \cdot 4} = \sqrt{1800} = 30\sqrt{2}$$

and the right answer is $\boxed{\textbf{(A) } 30\sqrt{2}}$

Problem 2

The triangle ABC has the sides 4, 4 and 6 and the triangle DEF has the sides 6, 6 and 4. Which of the following is the product of the areas of these triangles?

(A) $25\sqrt{13}$ **(B)** $42\sqrt{2}$ **(C)** $40\sqrt{7}$ **(D)** $22\sqrt{15}$ **(E)** $24\sqrt{14}$

Solution

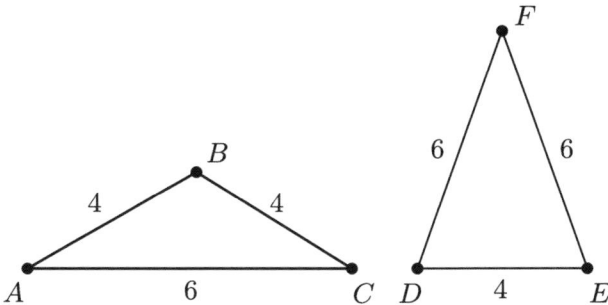

Notice that since all three sides of the triangles are given, then we can use the Heron's formula.

First, let us find the area S_1 of the triangle ABC

$$p = \frac{4+4+6}{2} = \frac{14}{2} = 7$$
$$p - a = 7 - 4 = 3$$
$$p - b = 7 - 4 = 3$$
$$p - c = 7 - 6 = 1$$

By Heron's formula the area of the triangle ABC is equal to

$$S_1 = \sqrt{7 \cdot 3 \cdot 3 \cdot 1} = \sqrt{63} = 3\sqrt{7}$$

Now, let us find the area of the triangle DEF

$$p = \frac{6+6+4}{2} = \frac{16}{2} = 8$$
$$p - a = 8 - 6 = 2$$
$$p - b = 8 - 6 = 2$$
$$p - c = 8 - 4 = 4$$

By Heron's formula the area S_2 of the triangle DEF is equal to

$$S_2 = \sqrt{8 \cdot 2 \cdot 2 \cdot 4} = \sqrt{128} = 8\sqrt{2}$$

Therefore, the product of the areas of the triangles ABC and DEF is

$$S_1 \cdot S_2 = 3\sqrt{7} \cdot 8\sqrt{2} = 24\sqrt{14}$$

and the right answer is $\boxed{\textbf{(E)} \; 24\sqrt{14}}$ [1]

Problem 3

The sides of the triangle ABC are given as $a = 5$, $b = 6$ and $c = 7$. What is the value of the inradius of the triangle ABC?

(A) $2\sqrt{3}$ **(B)** $3\sqrt{2}$ **(C)** $\sqrt{6}$ **(D)** $2\sqrt{2}$ **(E)** $\frac{2\sqrt{6}}{3}$

Solution

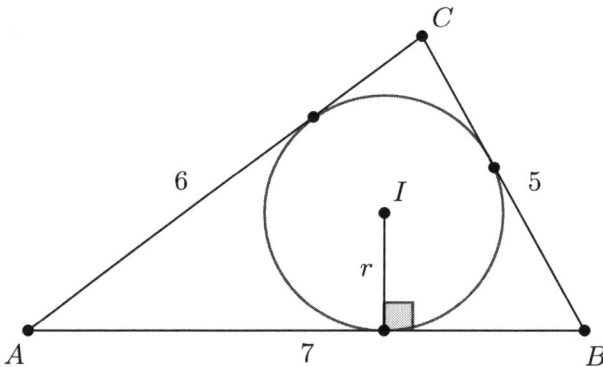

Notice that the area S of the triangle is equal to its semiperimeter p multiplied by its inradius r. From here we can find the inradius by dividing the area of the triangle by its semiperimeter

First, let us find the area of the triangle ABC using Heron's formula

$$p = \frac{5 + 6 + 7}{2} = \frac{18}{2} = 9$$
$$p - a = 9 - 5 = 4$$
$$p - b = 9 - 6 = 3$$
$$p - c = 9 - 7 = 2$$

By Heron's formula the area of the triangle ABC is equal to

$$S = \sqrt{9 \cdot 4 \cdot 3 \cdot 2} = \sqrt{216} = 6\sqrt{6}$$

---

[1]This problem can be also solved by finding the heights of the triangles using the Pythagorean Theorem discussed in detail in Chapter 52 "Pythagorean Theorem"

Therefore, the inradius r of the triangle ABC is

$$r = \frac{S}{p} = \frac{6\sqrt{6}}{9} = \frac{2\sqrt{6}}{3}$$

and the right answer is $\boxed{\textbf{(E)} \ \dfrac{2\sqrt{6}}{3}}$

CHAPTER 56

WHEN AREA HELPS

In many geometry problems we can use the concept of the **area** to find certain quantities, such as the inradius, the sides of the triangle or other segments, even though the area is not specifically mentioned in the problem. Usually this idea is employed using two or more formulas for the area.

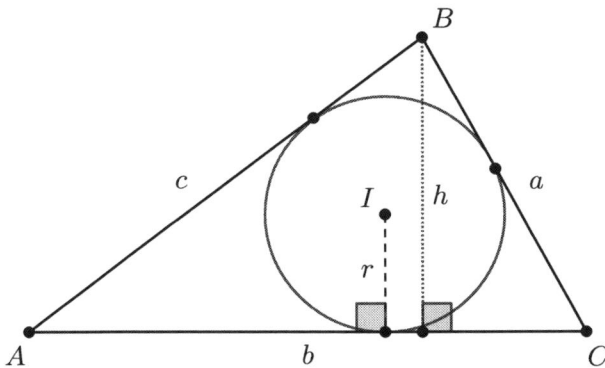

Let us assume that we are given a triangle ABC. Let $AC = b$, $AB = c$, $BC = a$. Let the height drawn from the vertex B be $BH = h$, r be the inradius and p the semiperimeter defined as

$$p = \frac{a + b + c}{2}$$

The area S of the triangle ABC can be found using the following formulas[1]

$$S = \frac{b \cdot h}{2}$$
$$S = \sqrt{p \cdot (p - a) \cdot (p - b) \cdot (p - c)}$$
$$S = p \cdot r$$

Let us consider several examples.

Problem 1

The sides of the triangle ABC are given as $AB = 5$, $BC = 5$ and $AC = 6$. What is the value of the inradius of the triangle ABC?

(A) 1   **(B)** $\frac{3}{2}$ **(C)** 2   **(D)** $\frac{5}{2}$ **(E)** 3

Solution

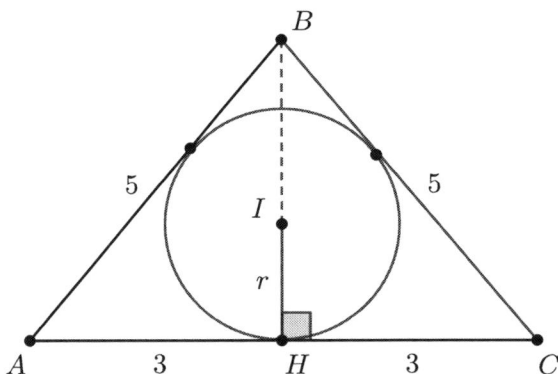

We will use the fact that the area S of the triangle is equal to its semiperimeter p multiplied by its inradius r. From here we can find the inradius by dividing the area of the triangle by its semiperimeter.

---

[1] The first two formulas are discussed in detail in Chapter 54 "Area of a Triangle" and Chapter 55 "Heron's Formula"

The semiperimeter of the triangle ABC is equal to

$$p = \frac{5 + 5 + 6}{2} = 8$$

Let us draw the altitude BH from the vertex B. Notice that ABC is isosceles and, therefore, BH is also the angle bisector and the median of the triangle ABC. This implies that the triangles ABH and CBH are congruent and represent a right triangle 3-4-5 with $BH = 4$. Therefore, the area[2] of the triangle ABC is equal to

$$S = \frac{6 \cdot 4}{2} = 12$$

and the inradius r of the triangle ABC is

$$r = \frac{S}{p} = \frac{12}{8} = \frac{3}{2}$$

Therefore, the right answer is $\boxed{\textbf{(B) } \dfrac{3}{2}}$

Problem 2

A quarter of a circle of radius R is inscribed in a right triangle with legs 3 and 4 as shown below.

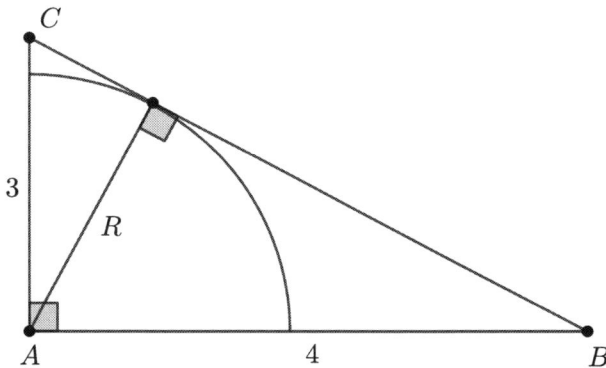

Find the value of R.

(A) $\frac{3}{4}$ (B) $\frac{4}{5}$ (C) $\frac{5}{12}$ (D) $\frac{12}{5}$ (E) $\frac{12}{13}$

---

[2]The area of the triangle ABC can be also found using Heron's Formula discussed in detail in Chapter 55 "Heron's Formula"

Solution

First let us find the area S of the triangle

$$S = \frac{3 \cdot 4}{2} = 6$$

Notice that the given triangle is a Pythagorean Triangle 3-4-5 and, therefore, the hypotenuse is equal to 5.

Let us draw the radius to the point of tangency with the hypotenuse. Since it is perpendicular to the hypotenuse, then it serves as a height of the triangle and, therefore, the area is equal to

$$\frac{5 \cdot R}{2} = 6$$

This equation can be solved by isolating R

$$\frac{5 \cdot R}{2} = 6$$
$$5R = 12$$
$$R = \frac{12}{5}$$

and the right answer is $\boxed{\textbf{(D) } \dfrac{12}{5}}$

Problem 3

Given a triangle MNK, such that $KN = 15$, $KM = 8$, and $\angle MKN = 90°$. A semicircle is tangent to the sides KM and MN and its diameter lies on the side KN. Find the radius of the semicircle.

(A) 4.4   **(B)** 4.5   **(C)** 4.6   **(D)** 4.7   **(E)** 4.8

Solution

Note that the triangle MNK is right, then the lengths of its sides form a Pythagorean triple 8-15-17 and, therefore, $MN = 17$.

Let us reflect the triangle MNK with respect to the line KN and let L be the image of the point M. Then the triangle LMN is isosceles and the given semicircle is its incircle. Now we can use the fact that the area S of the triangle is equal to its semiperimeter p multiplied by its inradius r. From here we can find the inradius by dividing the area of the triangle LMN by its semiperimeter.

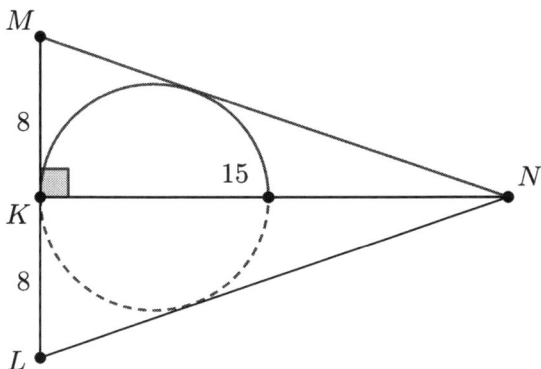

The semiperimeter of the triangle LMN is equal to

$$p = \frac{16 + 17 + 17}{2} = 25$$

The area S of the triangle LMN is

$$S = \frac{16 \cdot 15}{2} = 120$$

and the inradius r of the triangle LMN is

$$r = \frac{S}{p} = \frac{120}{25} = 4.8$$

and the right answer is $\boxed{\textbf{(E) } 4.8}$ [3]

---

[3] The area of the triangle LMN can also be found using Heron's Formula discussed in detail in Chapter 55 "Heron's Formula"

CHAPTER 57

SIMILAR TRIANGLES. PART 1

Triangles ABC and $A_1B_1C_1$ are called **similar** if their corresponding angles are congruent and their corresponding sides are proportional, i.e. it holds that

$$\angle ABC = \angle A_1B_1C_1$$
$$\angle BCA = \angle B_1C_1A_1$$
$$\angle CAB = \angle C_1A_1B_1$$

and

$$\frac{AB}{A_1B_1} = \frac{BC}{B_1C_1} = \frac{AC}{A_1C_1}$$

There are three basic similarity principles, namely: *Angle-Angle*, *Side-Angle-Side* and *Side-Side-Side*. In this chapter we will focus on the *Angle-Angle* similarity.

The *Angle-Angle* similarity states that if two pairs of corresponding angles are congruent, then the triangles are similar.

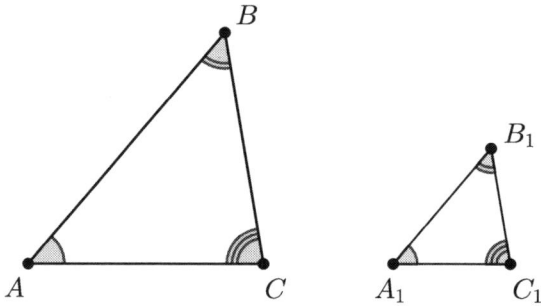

Problem 1

Point D is chosen on the side AB of the triangle ABC. It is given that $AC = 16$, $BC = 18$, $CD = 12$ and $\angle ACD = \angle ABC$. Which of the following equals to the length of the segment AD?

(A) 10   **(B)** 12   **(C)** $\frac{28}{3}$ **(D)** $\frac{32}{3}$ **(E)** $\frac{21}{2}$

Solution

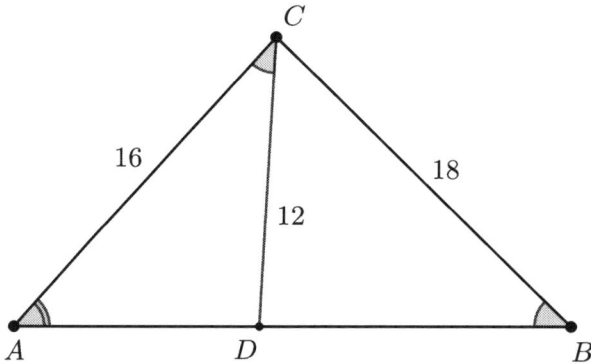

Note that since $\angle CAD = \angle BAC$ and $\angle ACD = \angle ABC$, then the triangles ACD and ABC are similar by *Angle-Angle* similarity.

Therefore, we have

$$\frac{AD}{AC} = \frac{12}{18}$$
$$\frac{AD}{16} = \frac{2}{3}$$
$$AD = \frac{32}{3}$$

and the right answer is $\boxed{\textbf{(D)}\ \dfrac{32}{3}}$

Problem 2

Let CD be the altitude in the right triangle ABC ($\angle ACB = 90°$), such that $AD = 54$ and $BD = 24$. Let the value to $AC - BC$ be expressed as $m\sqrt{n}$, where m and n are positive integers and n does not contain any perfect squares. Which of the following represents the value of $m + n$?

(A) 15   **(B)** 18   **(C)** 19   **(D)** 22   **(E)** 23

Solution

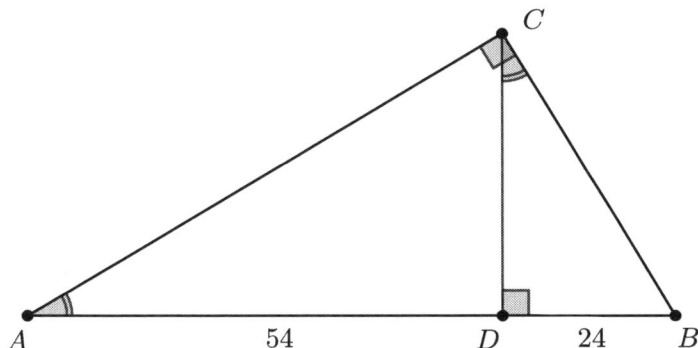

Note that if $\angle CAD = x$, then $\angle ACD = 90° - x$ and $\angle BCD = x$. Therefore, the triangles ACD and CBD are similar by *Angle-Angle* similarity. From here we have

$$\frac{AD}{CD} = \frac{CD}{BD}$$
$$\frac{54}{CD} = \frac{CD}{24}$$
$$CD^2 = 54 \cdot 24$$
$$CD^2 = 1296$$
$$CD = 36$$

Now we can apply the Pythagorean Theorem[1] to the triangles ACD and CBD to find the sides AC and BC respectively

---

[1] This theorem is discussed in detail in Chapter 52 "Pythagorean Theorem"

$$AC = \sqrt{AD^2 + CD^2} = \sqrt{(54)^2 + (36)^2} = 18\sqrt{13}$$
$$BC = \sqrt{BD^2 + CD^2} = \sqrt{(24)^2 + (36)^2} = 12\sqrt{13}$$

From here we have that

$$AC - BC = 18\sqrt{13} - 12\sqrt{13} = 6\sqrt{13}$$

Therefore, $m = 6$, $n = 13$, $m + n = 6 + 13 = 19$ and the right answer is $\boxed{\textbf{(C)} \ 19}$

Problem 3

The diagonal BD of an isosceles trapezoid $ABCD$ is perpendicular to the lateral side AB. The bases of the trapezoid are given as $BC = 20$ and $AD = 32$. Which of the following is closest to the value of the area of the trapezoid?

(A) 287   **(B)** 292   **(C)** 299   **(D)** 305   **(E)** 311

Solution

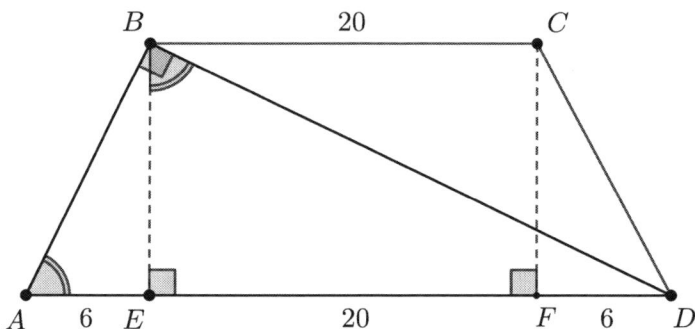

Let us start by dropping the altitudes BE and CF from the points B and C to the line AD respectively. Since the trapezoid $ABCD$ is isosceles, then the triangles ABE and DCF are congruent, and, therefore, $AE = FD$. So, we have

$$AD = AE + EF + FD$$
$$32 = AE + 20 + AE$$
$$32 = 2AE + 20$$
$$12 = 2AE$$
$$6 = AE$$

Note that if $\angle BAE = x$, then $\angle ABE = 90° - x$ and $\angle DBE = x$. Therefore, the triangles ABE and DBE are similar by *Angle-Angle* similarity. From here we have

$$\frac{AE}{BE} = \frac{ED}{BE}$$
$$\frac{6}{BE} = \frac{26}{BE}$$
$$BE^2 = 6 \cdot 26$$
$$BE^2 = 156$$
$$BE = 2\sqrt{39}$$

Now we can find the area of the trapezoid $ABCD$

$$\frac{(BC + AD) \cdot BE}{2} = \frac{(20 + 32) \cdot 2\sqrt{39}}{2} = 52\sqrt{39}$$

We can now estimate[2] the value of the area as

$$52\sqrt{39} > 52\sqrt{36} = 52 \cdot 6 = 312$$

and, therefore, the right answer is $\boxed{\textbf{(E) } 311}$

---

[2] You can find more problems that involve this technique in Chapter 25 "Estimations"

CHAPTER 58

SIMILAR TRIANGLES. PART 2

Triangles ABC and $A_1B_1C_1$ are called **similar** if their corresponding angles are congruent and their corresponding sides are proportional. There are three basic similarity principles, namely: *Angle-Angle*, *Side-Angle-Side* and *Side-Side-Side*. In this chapter we will focus on the *Side-Angle-Side* and *Side-Side-Side* similarities.

The *Side-Angle-Side* similarity states that if an angle of a triangle is congruent to an angle of another triangle and if the two pairs of sides adjacent to these angles are proportional, then the two triangles are similar.

The *Side-Side-Side* similarity states that if the three pairs of corresponding sides of two triangles are proportional, then the two triangles are similar.

Problem 1

In the triangle ABC the lengths of the sides AB and BC are equal 8 and 12 respectively. Point D is chosen on the side AC, such that $AD = 7$ and $DC = 9$. Find the length of the segment BD.

(A) 4   **(B)** 6   **(C)** 7   **(D)** 8   **(E)** 9

Solution

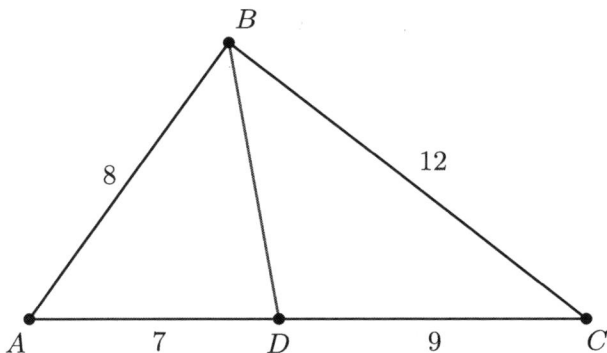

First, we can find the length of the side AC

$$AC = AD + DC = 7 + 9 = 16$$

Second, since

$$\frac{BC}{AC} = \frac{12}{16} = \frac{9}{12} = \frac{DC}{BC}$$

then the triangles BDC and ABC are similar by *Side-Angle-Side* similarity. This implies that

$$\frac{BD}{AB} = \frac{BC}{AC}$$
$$\frac{BD}{8} = \frac{12}{16}$$
$$BD = 6$$

and the right answer is $\boxed{\textbf{(B) } 6}$

Problem 2

The diagonals of the quadrilateral $ABCD$ intersect at the point O and the point E is belongs to the side AD. Given that $AB = 40$, $BC = 27$, $CD = 30$, $AE = ED = 24$ and $BO = OD = 18$. Which of the following represents the ratio $BE : CO$?

(A) $4 : 3$ **(B)** $3 : 2$ **(C)** $5 : 4$ **(D)** $7 : 5$ **(E)** $2 : 1$

Solution

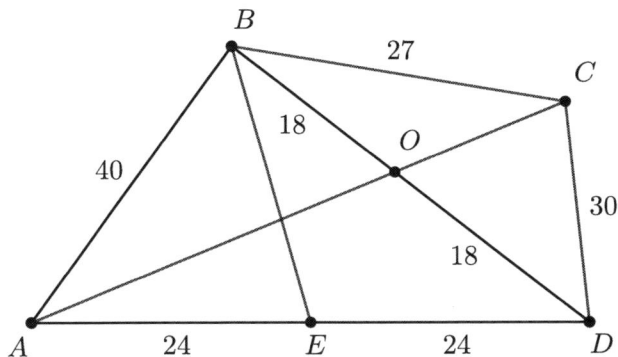

We can start by finding the lengths of the segments AD and BD

$$AD = AE + ED = 24 + 24 = 48$$
$$BD = BO + OD = 18 + 18 = 36$$

Notice that

$$\frac{AB}{CD} = \frac{40}{30} = \frac{4}{3}$$
$$\frac{AD}{BD} = \frac{48}{36} = \frac{4}{3}$$
$$\frac{BD}{BC} = \frac{36}{27} = \frac{4}{3}$$

then the triangles ABD and DCB are similar by *Side-Side-Side* similarity. However, since BE and CO are the medians in the triangles ABD and DCB respectively, then

$$\frac{BE}{CO} = \frac{4}{3}$$

and the right answer is $\boxed{\textbf{(A) } 4 : 3}$

Problem 3

Points D and E are chosen on the sides AB and BC of the triangle ABC and the point F is chosen on the segment DE. It appears that $EC = BD$, $AD - BE = 5$, $BD - BE = 2$, $EF = 3$, $AB = 15$. Find the length of the segment CN, where N is the intersection of BF and AC.

(A) 7 **(B)** $\frac{37}{5}$ **(C)** $\frac{29}{4}$ **(D)** $\frac{15}{2}$ **(E)** $\frac{22}{3}$

Solution

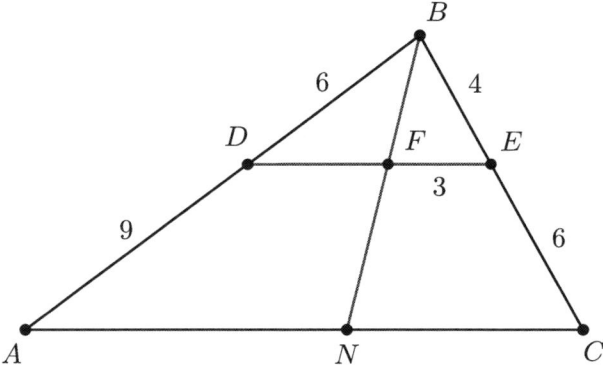

Let us assume that

$$AD = x$$
$$BD = y$$
$$BE = z$$

From here we have that $EC = BD = y$. The other conditions represent the following system of equations

$$\begin{cases} x - z = 5 \\ y - z = 2 \\ x + y = 15 \end{cases}$$

By adding the first two equations and subtracting the third equation we have

$$(x - z) + (y - z) - (x + y) = (5) + (2) - (15)$$
$$-2z = -8$$
$$z = 4$$

Substituting z into the first equation we get $x = 9$, and into the second equation we get $y = 6$. From here

$$AB = AD + DB = 9 + 6 = 15$$
$$BC = BE + EC = 4 + 6 = 10$$

Since

$$\frac{AB}{DB} = \frac{15}{6} = \frac{10}{4} = \frac{BC}{BE}$$

then the triangles ABC and DBE are similar by *Side-Angle-Side* similarity. This implies that $\angle DEB = \angle ACB$ and the triangles ABC and DBE are similar by

Angle-Angle similarity. Therefore

$$\frac{CN}{EF} = \frac{BC}{BE}$$
$$\frac{CN}{3} = \frac{10}{4}$$
$$CN = \frac{15}{2}$$

and the right answer is $\boxed{\textbf{(D)} \ \dfrac{15}{2}}$

CHAPTER 59

POWER OF A POINT

The **Power of a Point** P with respect to a circle with center O and radius r is defined as

$$PO^2 - r^2$$

If the point P is outside the circle, then its power is positive, and if the point P is inside the circle, then its power is negative. The power of the points on the circle are equal to zero.

The Power of a Point implies the following two theorems.

1. If the chords AB and CD intersect at the point X inside the circle, then

$$XA \cdot XB = XC \cdot XD$$

2. If the chords AB and CD intersect at the point X outside the circle, then

$$XA \cdot XB = XC \cdot XD = XE^2$$

where E is the point of tangency to the circle drawn from the point X.

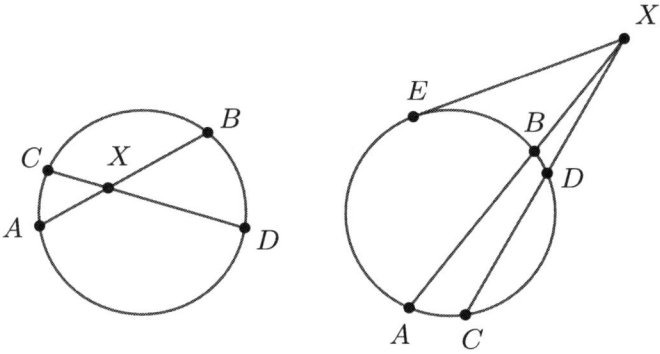

Let us now consider several examples.

Problem 1

Chords $AC = 15$ and $BD = 14$ intersect at the point O located inside the circle. Find the length of the segment OC if $OA - OB = 2$.

(A) $\frac{13}{3}$ (B) $\frac{14}{3}$ (C) $\frac{9}{2}$ (D) $\frac{15}{4}$ (E) $\frac{21}{5}$

Solution

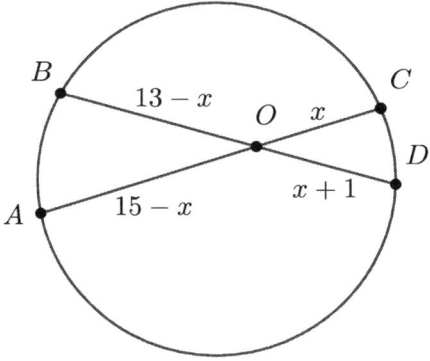

Let us put $OC = x$. Then $OA = 15 - x$ and we can find OB as follows

$$OA - OB = 2$$
$$(15 - x) - OB = 2$$
$$13 - x = OB$$

We can also find OD from

$$OB + OD = BD$$
$$(13 - x) + OD = 14$$
$$OD = x + 1$$

Now we can find x from the Power of a Point

$$OA \cdot OC = OB \cdot OD$$
$$(13 - x) \cdot (x + 1) = x \cdot (15 - x)$$
$$13x + 13 - x^2 - x = 15x - x^2$$
$$-3x = -13$$
$$x = \frac{13}{3}$$

and the right answer is $\boxed{\textbf{(A)}\ \dfrac{13}{3}}$

Problem 2

Given a circle that passes through the points $A(2, 5)$ and $B(7, 15)$. Find the length of the tangent line drawn to the circle from the point $C(-4, -7)$.

(A) $\sqrt{318}$ **(B)** $\sqrt{330}$ **(C)** $\sqrt{346}$ **(D)** $\sqrt{353}$ **(E)** $\sqrt{359}$

Solution

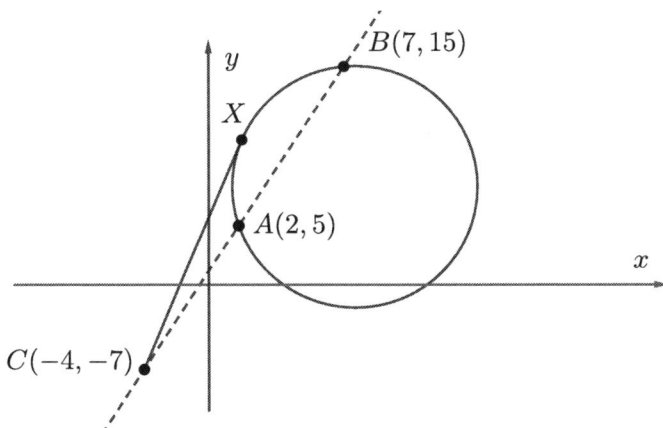

Start by noticing that all the mentioned points are located on the line $y = 2x + 1$. Indeed

$$5 = 2(2) + 1$$
$$15 = 2(7) + 1$$
$$-7 = 2(-4) + 1$$

Therefore, the line containing the points A and B represents a secant drawn from the point C.

Let us now find the lengths of the segments CA and CB from the distance formula

$$CA = \sqrt{((-4) - (2))^2 + ((-7) - (5))^2} = \sqrt{180} = 6\sqrt{5}$$
$$CB = \sqrt{((-4) - (7))^2 + ((-7) - (15))^2} = \sqrt{605} = 11\sqrt{5}$$

We can find the length of the tangent line drawn from the point C using the Power of a Point. If X is the point of tangency, then we have

$$CX^2 = CA \cdot CB$$
$$CX^2 = 6\sqrt{5} \cdot 11\sqrt{5}$$
$$CX^2 = 330$$
$$CX = \sqrt{330}$$

and the right answer is $\boxed{\textbf{(C)} \ \sqrt{330}}$

Problem 3

In the right triangle ABC ($\angle ABC = 90°$) the lengths of the legs AB and BC are equal to 55 and 48 respectively. Point D is the point of tangency of the incircle of the triangle ABC and the side BC. Let E be the point of intersection of the segment AD and the incircle. Find the length of the segment AE.

(A) $\frac{29\sqrt{130}}{10}$ **(B)** $\frac{30\sqrt{130}}{13}$ **(C)** $\frac{31\sqrt{130}}{10}$ **(D)** $\frac{32\sqrt{130}}{13}$ **(E)** $\frac{33\sqrt{130}}{13}$

Solution

We will start by applying the Pythagorean Theorem[1] to the triangle ABC

$$AC = \sqrt{AB^2 + BC^2} = \sqrt{(55)^2 + (48)^2} = \sqrt{5329} = 73$$

---

[1] This theorem is discussed in detail in Chapter 52 "Pythagorean Theorem"

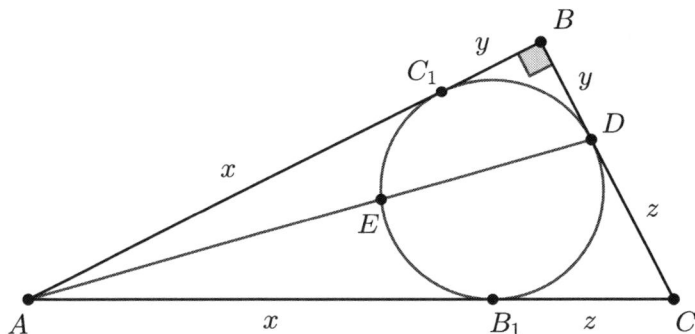

Now let B_1 and C_1 be the points of tangency of the incircle of the triangle ABC and the sides AC and AB respectively. Let us put

$$AB_1 = AC_1 = x$$
$$BC_1 = BD = y$$
$$CB_1 = CD = z$$

Then we have the following system of equations

$$\begin{cases} x + y = 55 \\ y + z = 48 \\ z + x = 73 \end{cases}$$

By adding the three equations we have

$$(x + y) + (y + z) + (z + x) = 55 + 48 + 73$$
$$2(x + +y + z) = 176$$
$$x + y + z = 88$$

From here $x = 40$, $y = 15$, $z = 33$.

Now we will apply the Pythagorean Theorem to the triangle ABD

$$AD = \sqrt{AB^2 + BD^2} = \sqrt{(55)^2 + (15)^2} = \sqrt{3250} = 5\sqrt{130}$$

By the Power of a Point we have

$$AC_1^2 = AE \cdot AD$$
$$40^2 = AE \cdot 5\sqrt{130}$$
$$\frac{32\sqrt{130}}{13} = AE$$

and the right answer is $\boxed{\textbf{(D)}\ \dfrac{32\sqrt{130}}{13}}$

CHAPTER 60

MASS POINTS

Mass Points is a technique that is very useful for the problems involving ratios of segments and intersecting lines.

Let us assume that we are given a point B. A *mass point* is an ordered pair of the form

$$(m_B, B)$$

where m_B is a positive real number called *mass*. For example, if the mass of the point B is equal to 5, then the mass point will be $(5, B)$ or simply $5B$.

The addition of two mass points (m_A, A) and (m_B, B) is a new mass point (m_X, X), such that the following two conditions are satisfied

- the mass m_X equals to the sum of the masses of (m_A, A) and (m_B, B)

$$m_X = m_A + m_B$$

- the point X lies on the segment AB, such that

$$m_A \cdot AX = m_B \cdot XB$$

For example, if the mass of the point A is equal to 3 and the mass of the point B is equal to 1, then the sum of these mass points is a new mass point (X, m_X) with the mass $m_X = 3 + 1 = 4$ located on the segment AB, such that $AX : XB = 1 : 3$.

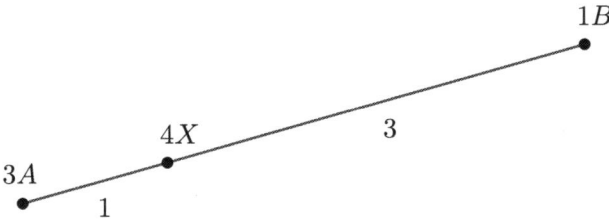

The lines that pass through exactly one vertex of the triangle are called *cevians*. The lines that do not pass through any of the vertices of the triangle are called *transversals*. For example, the line BD is a *cevian* of the triangle ABC and the line HI is the *transversal* of the triangle EFG.

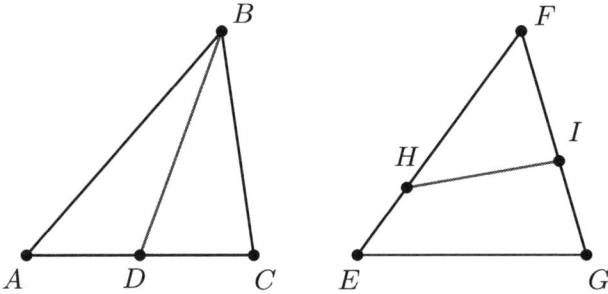

The mass point calculations are different depending on which of the two lines we work with in a particular problem: *cevians* or *transversal*.

If you work with *cevians*, then follow these two principles:

- the foot of a *cevians* is always the sum of the two vertices it does not pass through

- the point of concurrency of the *cevians* is always the sum of the vertex and the foot of the *cevian*

If you work with *transversals*, then follow these two principles:

- the vertices that belong to both sides intersected by the *transversal* should be assigned a *split mass*, i.e. three numbers instead of one

- the *split mass* consists of three masses: one mass that is used for one side, one mass that is used for another side, and the sum of these masses that is used for its *cevians*

Problem 1

Points D, E and F belong to the sides BC, AC and AB of the triangle ABC, such that the lines AD, BE and CF are concurrent at O and $AF : FB = 2 : 3$ and $AE : EC = 3 : 4$. Find the ratio $BD : DC$.

(A) $4 : 3$ (B) $3 : 4$ (C) $9 : 8$ (D) $8 : 9$ (E) $1 : 1$

Solution

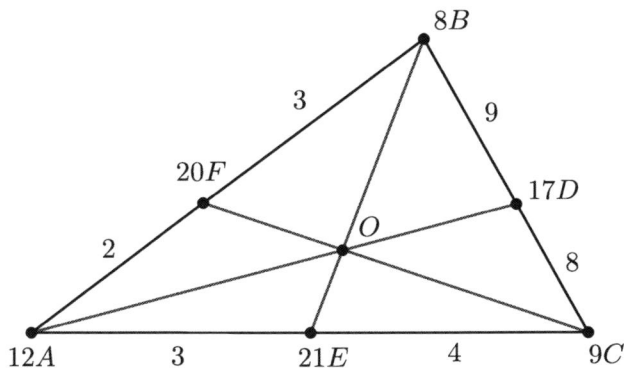

Notice that in this problem we work with the *cevians*.

Let us assign the mass of the point A to be $m_A = 12$. The mass of the point B can be found from the segment AB and the equation

$$m_A \cdot AF = m_B \cdot FB$$

which becomes

$$12 \cdot 2 = m_B \cdot 3$$
$$24 = 3m_B$$
$$8 = m_B$$

The mass of the point C can be found from the segment AC and the equation

$$m_A \cdot AE = m_C \cdot EC$$

which becomes

$$12 \cdot 3 = m_C \cdot 4$$
$$36 = 4m_C$$
$$9 = m_C$$

Since the lines AD, BE and CF are concurrent, then

$$m_B \cdot BD = m_C \cdot DC$$

which is equivalent to

$$\frac{BD}{CD} = \frac{m_C}{m_B} = \frac{9}{8}$$

and the right answer is $\boxed{\textbf{(C)} \ 9 : 8}$

Problem 2

Points D, E and F belong to the sides BC, AC and AB of the triangle ABC, such that the lines AD, BE and CF are concurrent at O and $AF : FB = 1 : 5$ and $AE : EC = 1 : 5$. Find the ratio $AO : OD$.

(A) $1 : 2$ **(B)** $2 : 1$ **(C)** $1 : 1$ **(D)** $5 : 2$ **(E)** $2 : 5$

Solution

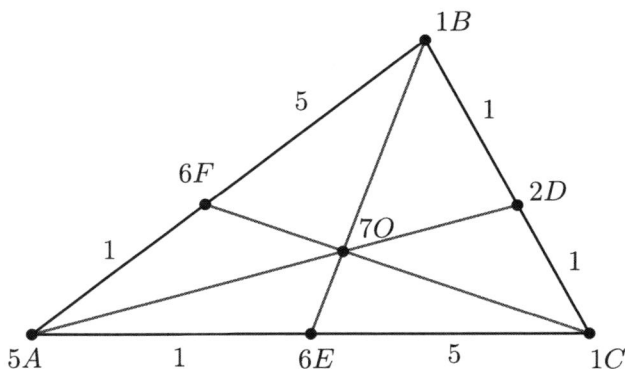

Notice that in this problem we work with the *cevians*.

Let us assign the mass of the point A to be $m_A = 5$. The mass of the point B can be found from the segment AB and the equation

$$m_A \cdot AF = m_B \cdot FB$$

which becomes

$$5 \cdot 1 = m_B \cdot 5$$
$$1 = m_B$$

The mass of the point C can be found from the segment AC and the equation

$$m_A \cdot AE = m_C \cdot EC$$

which becomes

$$5 \cdot 1 = m_C \cdot 1$$
$$5 = m_C$$

Since the lines AD, BE and CF are concurrent, then the mass of the point D is equal to

$$m_D = m_B + m_C = 2$$

Therefore, from the segment AD we have

$$m_A \cdot AO = m_D \cdot OD$$

which is equivalent to

$$\frac{AO}{OD} = \frac{m_D}{m_A} = \frac{2}{5}$$

and the right answer is $\boxed{\textbf{(E)}\ 2:5}$

Problem 3

Points K, M and N are chosen on the sides AB, BC and AC of the triangle ABC, such that $AK : KB = 4 : 3$, $BM : MC = 1 : 1$, $AN : NC = 3 : 2$. The line BN intersects the line KM at the point L. Find the ratio $KL : LM$.

(A) $4:7$ **(B)** $7:4$ **(C)** $1:1$ **(D)** $9:7$ **(E)** $7:9$

Solution

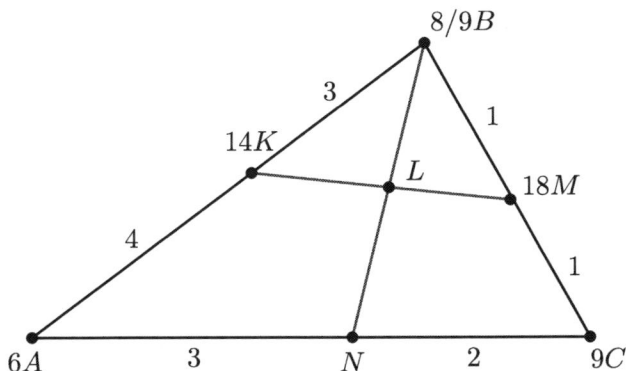

Notice that in this problem we work with a *transversal*.

Let us find the mass of the point B in two ways. Let us assign the mass of the point A to be $m_A = 6$. The first way to find the mass of the point B will be from the segment AB and the equation

$$m_A \cdot AK = m_B \cdot KB$$

which becomes

$$6 \cdot 4 = m_B \cdot 3$$
$$24 = 3m_B$$
$$8 = m_B$$

The second way to find the mass of the point B will be from the segment AC and the mass of the point C. The mass of the point C can be found from the segment AC and the equation

$$m_A \cdot AN = m_C \cdot NC$$

which becomes

$$6 \cdot 3 = m_C \cdot 2$$
$$18 = 2m_C$$
$$9 = m_C$$

We can now find the mass of the point B from the segment BC and the equation

$$m_B \cdot BM = m_C \cdot MC$$

which becomes

$$m_B \cdot 1 = 9 \cdot 1$$
$$m_B = 9$$

Therefore, the point B will have a *split mass* of $8/9$.

The mass of the point K is equal to

$$m_K = m_A + m_B = 6 + 8 = 14$$

The mass of the point M is equal to

$$m_M = 9 + 9 = 18$$

Therefore, from the segment KM we have

$$m_K \cdot KL = m_M \cdot LM$$

which is equivalent to

$$\frac{KL}{LM} = \frac{m_M}{m_K} = \frac{18}{14} = \frac{9}{7}$$

and the right answer is $\boxed{\textbf{(D)} \ 9 : 7}$

Made in United States
Troutdale, OR
09/25/2024